Modeling and Simulation in Ecotoxicology with Applications in MATLAB® and Simulink®

Modeling and Simulation in Ecotoxicology with Applications in MATLAB® and Simulink®

Kenneth R. Dixon

CRC Press
Taylor & Francis Group
Boca Raton London New York

CRC Press is an imprint of the
Taylor & Francis Group, an **informa** business

CRC Press
Taylor & Francis Group
6000 Broken Sound Parkway NW, Suite 300
Boca Raton, FL 33487-2742

First issued in paperback 2019

ISBN-13: 978-1-4398-5517-1 (hbk)
ISBN-13: 978-1-138-37448-5 (pbk)

Library of Congress Cataloging-in-Publication Data

Dixon, Kenneth R.
    Modeling and simulation in ecotoxicology with applications in MATLAB and Simulink / Kenneth R. Dixon.
        p. ; cm.
    Includes bibliographical references and index.
    ISBN 978-1-4398-5517-1 (hardcover : alk. paper)
    1. Environmental toxicology--Data processing. I. Title.
    [DNLM: 1. MATLAB. 2. SIMULINK. 3. Ecotoxicology--methods. 4. Computer Simulation. 5. Mathematical Computing. 6. Software. WA 670]

RA1226.D59 2012
615.9'020113--dc23

2011013518

Visit the Taylor & Francis Web site at
http://www.taylorandfrancis.com

and the CRC Press Web site at
http://www.crcpress.com

In Memory of Dorothy Phipps and Mildred Wanninger

Contents

Preface

This book is about the role of modeling and simulation in environmental toxicology. It covers the steps in modeling and simulation from problem conception to validation and simulation analysis. Examples of mathematical functions and simulations are presented using the MATLAB® and Simulink® programming languages. We have proposed including this text in the MATLAB book series. The main themes are how to develop mathematical models and run computer simulations of the effects of toxic agents on biological and ecological processes using MATLAB software. This book is designed as a textbook for advanced undergraduate and graduate courses in the field of environmental toxicology. We try to present the modeling in a rigorous, yet easy-to-understand, manner. The book can be used in courses with students who have little or no experience in modeling. We also include MATLAB *m*-files or Simulink block diagrams for most examples in the text and on a CD. Although the methodology emphasizes environmental toxicology, it can be applied rather easily to other biological fields.

Chapter 1 is intended to introduce the student to the use of models in general and environmental toxicology in particular. It describes how modeling and simulation play a role in the broader context of ecological research. The introduction does not include equations to avoid apprehension on the part of students with little quantitative experience. It also sets the stage for the types of models covered in the text. Chapter 2 presents the general principles of modeling and simulation based upon existing literature and the author's forty years of experience. The steps in modeling and simulation are described, including parameter estimation, experimental design, analysis of simulation experiments, and validation, which are explored more fully in later chapters. Chapter 3 describes the foundation for our modeling and simulation, which are the programming languages, MATLAB and Simulink. These are widely used in a variety of disciplines because of their wide range of functions and a history of software verification. We present a brief introduction to both MATLAB and Simulink and then cover the functions used in subsequent chapters. Chapter 4 introduces stochastic modeling where variability and uncertainty are acknowledged by making parameters random variables. Parameter values can be drawn from a wide range of probability distributions as built-in MATLAB functions. We also describe probabilistic models such as Markov chains. The methodology of Monte Carlo simulation also is described. Chapter 5 describes toxicological processes from the level of the individual organism. We include worked examples of process models in either MATLAB or Simulink or both. The model descriptions include MATLAB code (m-files) or Simulink block diagrams.

Chapter 6 describes toxicological processes at the level of populations, communities, and ecosystems. Worked examples of population and ecosystem models in MATLAB and Simulink are included. Chapter 7 presents parameter estimation using least squares regression methods. The advantages and disadvantages of linear and nonlinear regression are discussed. Examples of both techniques are presented using MATLAB. Chapter 8 presents the design of simulation experiments similar to the experimental design applied to laboratory or field experiments. The emphasis is on identifying significant parameters and reducing the number of simulation experiments using fractional factorial designs. Examples using MATLAB are presented, including response surface methods. Chapter 9 describes several methods of postsimulation analysis, including stability analysis and sensitivity analysis. Stability measures are presented for the transient response to a unit impulse function. Sensitivity analysis is described as the relative change in state variables to changes in parameter values. Examples in MATLAB are presented. Chapter 10 presents the complex and controversial topic of model validation. We present a consensus view but discuss the different levels of validation and how these are related to modeling purpose. Examples of statistical methods of validation are included.

Chapter 11 presents a case study of a model developed to assess the relative risk of mortality following exposure to insecticides in different avian species. Many of the toxicological processes described in Chapter 5 are included in the model. The steps in the modeling and simulation are described for this model. Chapter 12 is a case study designed to explore the role of diving behavior on the inhalation and distribution of naphthalene in bottlenose dolphins. The model is an example of physiologically based toxicokinetic models. The case study in Chapter 13 looks at the dynamics of mercury in Daphnia that are exposed to simulated thermal plumes from a hypothetical power plant cooling system. Differences in ambient water temperature and cooling conditions are explored in the simulations.

Kenneth R. Dixon
Texas Tech University
Lubbock, Texas

MATLAB® and Simulink® are registered trademarks of The MathWorks, Inc. For product information, please contact:

The MathWorks, Inc.
3 Apple Hill Drive
Natick, MA 01760-2098 USA
Tel: 508-647-7000
Fax: 508-647-7011
E-mail: info@mathworks.com
Web: www.mathworks.com

Acknowledgments

As I argue in Chapter 1, modeling and simulation require at least a modicum of math and computer programming skills as well as a strong foundation in the basic science discipline (in this text, its ecotoxicology). It is unusual for someone to have all those skills to the degree necessary. In my case, I have spent time in both the quantitative and the natural science camps and can attest to the fact that these are not easy disciplines to master. Also, I have had the good fortune to have been associated with, and mentored by, some of the best, who have imparted some of their wisdom in these matters.

Early in my years in graduate school, I was fortunate to take a course in systems ecology taught by one of its founders, George M. Van Dyne. This course piqued my interest in applying systems analysis techniques to environmental issues. I was then encouraged to develop the necessary foundations in math, statistics, and systems engineering by my advisor George W. Cornwell at the University of Florida. In this effort, I was supported with patience and encouragement by Richard L. Patterson. It was Dr. Patterson who invited me to the University of Michigan to pursue my Ph.D. in natural resource systems in the School of Natural Resources and the Environment. It was he, and my advisor at SNRE, John Kadlec, who provided unlimited encouragement and guidance, which gave me the confidence to continue the modeling and simulation aspects of my career.

As a member of the Systems Ecology Group in the Environmental Sciences Division at Oak Ridge National Laboratory, I was inspired to further advance my skills in applying systems analysis to environmental problems. Members of this group routinely shared ideas and collaborated on projects involving modeling and simulation. Particularly helpful in this regard were Don DeAngelis, Hank Schugart, Robert O'Neill, J. B. Mankin, Bob Goldstein, Jerry Olsen, and M. R. Patterson.

At the University of Maryland, Joseph A. Chapman encouraged my development of a graduate course that became the precursor of the course for which this book was written. It was also his encouragement that led to several quantitative methods in the analysis of radio-telemetry data. Thanks go to Ron Kendall, director of The Institute of Environmental and Human Health, for providing support for the completion of this book. I also thank Lenwood Hall, John Huckabee, Rami Naddy, and Jennifer Gottschalk Walters for the use of their data.

I have also been fortunate to have many outstanding graduate students who have not only developed their own skills in modeling and simulation, but improved mine as well. These include Eric Albers, Randy Apodaca, Pinar Dogru, Doug Florian, William Henriques, Dan Jacobs, Min Lian, Smita Sathe, Lori Sheeler, and Fred Snyder. Eric Albers and Lori Sheeler developed some of the MATLAB programs in the book. I have modified most of them and any programming errors are my own. Everyone should have a colleague to provide the expertise that one is lacking. In my case, that person was Sam Anderson, who taught me most of what I have learned about programming in MATLAB and Simulink. I thank all the technical support staff at The MathWorks, who too often had to correct my programming errors or provide the solutions to programming problems.

Finally, I gratefully acknowledge the support of my family—my wife, Sheila, and Buster, Cybele, and Beauregard, without whose loving support over the past fifteen years, the completion of this book would not have been possible.

Kenneth R. Dixon
Texas Tech University
Lubbock, Texas

Supplementary Resources Disclaimer

Additional resources were previously made available for this title on CD. However, as CD has become a less accessible format, all resources have been moved to a more convenient online download option.

You can find these resources available here: www.routledge.com/9781439855171

Please note: Where this title mentions the associated disc, please use the downloadable resources instead.

About the Author

Kenneth R. Dixon is professor in the Department of Environmental Toxicology and The Institute of Environmental and Human Health at Texas Tech University. He received his B.S. degree in forestry from the University of Florida in 1964 and his M.S. in forestry in 1968, also from the University of Florida, specializing in statistics and systems engineering. From 1968 to 1971, Dr. Dixon worked as a biometrician in the Institute of Statistics at North Carolina State University. In 1974, he received his Ph.D. from the School of Natural Resources at the University of Michigan. After graduating, he took a postdoctoral position as an ecologist and modeler at Oak Ridge National Laboratory. His research primarily involved modeling the impact of heavy metals on both aquatic and terrestrial ecosystems. Additional activities included the environmental impact assessment of nuclear power plants. In 1976, Ken joined the University of Maryland Appalachian Environmental Laboratory, where he taught courses in quantitative methods in wildlife management. His research included furbearer population dynamics, wildlife as monitors of environmental contamination, and spatial models of wildlife behavior.

From 1984 to 1992, Dr. Dixon headed the Wildlife Research Program in the Washington Department of Wildlife. Research activities primarily involved the use of radio telemetry, remote sensing, and geographic information systems to determine wildlife habitat requirements. Species studied included the peregrine falcon, spotted owl, pygmy rabbit, harbor seal, California sea lion, grizzly bear, gray wolf, elk, and mule deer. In 1985 he planned and implemented the department's geographic information system, which he then supervised until leaving the department in 1992. Under Ken's supervision, a statewide wildlife habitat mapping project was initiated. His home range analysis program was written into Arc/GIS to facilitate the calculation of habitat selection. Ken also assisted the U.S. Fish and Wildlife Service in implementing the Washington State Gap Analysis project.

Dr. Dixon was associate professor in the Department of Environmental Toxicology and the Institute of Wildlife and Environmental Toxicology at Clemson University from 1992 to 1997. His research included developing and applying computer simulation models to predict the movement and effects of toxic chemicals on wildlife populations and the environment, including the spatial distribution of toxicants and effects at ecosystem, landscape, and regional scales using geographic information systems. An example of research in this area was a study of topographic, soil, and weather parameters affecting the runoff of pesticides into farm ponds in Iowa and Illinois.

His current research interests include developing and applying computer simulation models to predict the movement of toxic chemicals in the environment and their effects on human and wildlife populations. Dr. Dixon also studies the spatial distribution of toxicants and effects at ecosystem, landscape, and regional scales by integrating models with geographic information systems. Current research projects include developing terrestrial food-chain models to predict the uptake and effects of pesticides, perchlorate, and explosives, and developing spatial models of the spread of infectious diseases, and a real-time model of exposure and effects of atmospheric pollutants. Dr. Dixon has taught courses in modeling, geographic information systems, ecosystems analysis, biometry, and wildlife management.

1 Introduction

In a text on modeling and simulation, it is important to define these terms as there is a wide range of usage for both. In fact, the two terms are often confused or used interchangeably. Some authors have used the term *simulation model*, which makes it necessary to define both model and simulation to distinguish simulation models from other types of models. We make a distinction between modeling and simulation to clarify our discussion of both terms. A *model* can be defined as a simplified abstract of a real-world object. For our purposes, *modeling* has the following quantitative definition:

Definition 1.1: *Modeling* is the process of using abstract mathematical representations of a system for analyzing or studying the relationships among the components of the system. ∎

Although related to modeling, simulation has more to do with what we do with a model after it is constructed. A formal definition of simulation appears as Definition 1.2.

Definition 1.2: *Simulation* is the process of using a computer to exercise a model for the purpose of mimicking the behavior of a real system. ∎

Having defined modeling, we could ask, "Why do modeling?" Everyone uses an implicit or "mental" model to consider the future consequences of today's actions (Gentner and Stevens 1983, Morgan et al. 2002). For example, consider the impacts of the release of a toxic chemical into the environment. One can imagine where the chemical may be transported, what plants and animals may be exposed, and what the effects of that exposure are. This mental model can be thought of as the first step in a more explicit model. The question then is not whether we should do modeling, but what kind of model we will use. An explicit model, one based upon quantitative descriptions and data, provides a more structured approach than a mental model. As put forward by the late wildlife statistician, Douglas Chapman, "the use of mathematical language and mathematical models introduces and forces clarity of ideas that may be otherwise lacking" (Chapman 1971, 429).

Although modeling and simulation are relatively new to environmental toxicology, there is a long and productive history of modeling in other biological disciplines, including population biology, ecology, and wildlife management. These integral disciplines usually are described as either mathematical biology (emphasizing biology), or biomathematics (emphasizing mathematics). Other biological disciplines that have a history of incorporating the use of models include biometrics, cybernetics, and systems ecology.

1.1 THEORIES UNDERLYING PREDICTIVE MODELS

The theories behind different types of models are based on a combination of biology, toxicology, mathematics, and statistics (Figure 1.1). Except for the time series models, which truly are naïve models (i.e., they contain no biological information), all of the models contain some biological or toxicological process and mathematical or statistical structure. Those models on the left side of Figure 1.1 emphasize the biological mechanism and those on the right side emphasize the statistical structure. Typical mechanistic models are time dependent and are described by difference

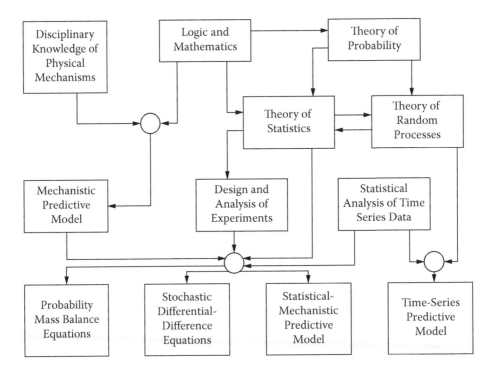

FIGURE 1.1   Theories underlying predictive models (R. L. Patterson, personal communication).

or differential equations. Statistical models typically are static and include regression, principle component, or Bayesian structures. Statistical-mechanistic models usually have a difference or differential equation structure but rely on statistical analysis of experimental data for parameterization. Stochastic differential or difference equations can take several forms. For our purposes, the models are referred to as *stochastic* or *probabilistic* in that some (or all) of the parameters in the model are random variables. The emphasis in this text is on predictive models using either stochastic differential or stochastic difference equations.

There are advantages and disadvantages to mechanistic models. One advantage is that the parameters in a mechanistic model usually have real-world counterparts. These parameters then are amenable to sensitivity analyses to determine the relative impact each parameter has on the process. Parameters in statistical models, however, have no biological meaning attached to them. Sensitivity analysis of these model parameters would not provide any added knowledge about the process. Largely because mechanistic models reflect physical and chemical dynamics of the process being modeled, they can predict future behavior of the process more accurately than statistical models can. Disadvantages of the mechanistic modeling approach are the relatively high cost and the amount of time required to conduct the necessary experiments to estimate the model parameters.

1.3   REASONS FOR MODELING AND SIMULATION

There are many reasons given for doing modeling and simulation. Generally, most reasons for modeling and simulation involve some aspect of prediction designed to provide insight or greater understanding of system behavior. Particularly, for our interest, we can say that the objective of modeling and simulation is to study the behavior of systems impacted by toxic compounds. Within that general purpose, there are several reasons to perform simulation studies. The reasons in the following sections provide a synthesis of most compilations (Hermann 1967, Martin 1968, Epstein 2008).

1.2.1 ALTERNATIVES AND THEIR CONSEQUENCES

A model may be used to predict the consequences of perturbations on a given system. A particular disturbance can be built into a model and the resulting effect on the system can be predicted. A modeler is not limited, however, to the real-world conditions reflected in model development. There are parameters that cannot be controlled very easily in the real-world system but can be controlled in the model. Once a model has been developed, parameters in the model can be changed to predict the effects of the changes on the real-world system. For example, a model may be designed to predict the impact of cleaning up a contaminated waste site. The cleanup would reduce the level of contamination and wildlife exposure but also could destroy important wildlife habitat. The model could predict the effects on habitats and animal body burdens for alternative levels of contaminant removal. Analysis of the results could be used to determine the optimal level of cleanup. This type of simulation experiment also can provide insight into system behavior in the real world. Simulation experiments, combined with laboratory and field studies, should provide greater understanding of the real-world system than either type of experiment conducted alone.

1.2.2 RELATIVE PREDICTIVE ABILITY

The purpose of modeling in this case involves predicting system behavior in a relative sense as opposed to an absolute sense. That is, the model is used to generate outcomes of effects of policy decisions as before, but not to predict system behavior accurately and with precision. Rather, the model is used as one tool for making predictions. Additional predictions can be made by other types of models (see Figure 1.1), including implicit models. Taking the previous example of cleanup of a contaminated waste site, a decision maker may have alternative models available such as statistical models. He or she also may have considerable experience in waste site cleanup and as a result have an intuitive understanding of the likely effects. The new model then would be used to provide additional *weight of evidence* for a particular management decision.

1.2.3 INSTRUCTION

Models have been used for teaching students about real-world systems. Students who have had courses that use simulation as a teaching tool often learn more than students who have taken comparable courses without a modeling component (Rieber et al. 2004, Windschitl and Andre 1998). Models that are used in instruction need to be as representative of real-world systems as possible to assure that students do not incorrectly draw conclusions about how systems function or what impact toxic substances will have on those systems. This will require that parameters with real-world counterparts have realistic values and in the case of parameters that vary during a simulation, they should be sampled from appropriate probability distributions. Simplified models that do not completely mimic real-world systems still can be useful teaching tools if the purpose is to explore theoretical concepts rather than to understand system behavior.

1.2.4 HYPOTHESIS AND THEORY CONSTRUCTION

Models can be used not only to make predictions about a system that functions within a known range of parameter values; a model also can be used to ask "what if" questions about system behavior. Simulations with "hypothetical" parameter values or changing relationships among variables in the model can lead to hypotheses about how a system functions. These hypotheses then can be tested by conducting laboratory or field experiments. After a number of successful predictions, a new theory of system behavior might be constructed. Sometimes a model may predict an outcome that runs counter to conventional wisdom or is seen as questioning authority. If the model is found to be accurate, a new hypothesis or theory could be developed. As an example, it is generally assumed that plants will accumulate atmospheric pollutants until the plant concentration reaches equilibrium equal to the atmospheric concentration. A mechanistic model of the uptake of atmospheric pollutants, used to predict the accumulation of radioactive carbon, ^{14}C, showed an accumulation much higher than the atmospheric concentration (Killough et al. 1976, Dixon and Murphy 1979). The reason for the difference was that the assumption of equilibrium failed to consider that carbon is fixed by the plant during photosynthesis.

1.2.5 NONEXISTENT UNIVERSES

We have, so far, discussed the uses of models of existing systems, although we might simulate conditions not observed. We also might wish to make predictions about behavior of systems that currently do not exist or may exist in the future. Models of systems that do not exist can be based upon models of existing systems if the two systems are known not to differ significantly. This requires the assumption that model parameters accurately represent comparable variables in the nonexistent system. Systems that we may consider creating in the future could include construction of a wetland for nutrient removal and applying new methods of remediating contaminated sites.

1.2.6 COST

Simulations can reduce both the monetary and time expenditures of experiments. If we are interested in knowing the effects of a toxicant on an ecosystem, one approach would be to introduce the contaminant into the system and then measure the effects. The resulting cost, however, could be unacceptable, both in terms of damage to the system and the cost of conducting the experiment. Simulation allows the experiment to be conducted a priori on the computer without the risk of environmental damage. Simulation experiments can be run in a fraction of the time it takes to run laboratory or field experiments. Most computer simulations take only a few seconds or minutes on modern computers, whereas laboratory or field experiments can take weeks or months.

1.2.7 PLANNING AND MANAGEMENT DECISION AID

Planners and managers need the best information available to make optimal decisions. Whether the question being addressed concerns the impact of introducing a new pollutant into a system or the best method of removing existing contaminants from a system, simulation can aid in providing information on the long-term effects of alternative decisions. A particular management decision can be built into a model and the resulting effect on the system can be predicted. A manager or decision maker may formulate alternative policies to be simulated. They can then evaluate and compare the consequences of their various management scenarios by analyzing the results of the model simulations. For example, a scientific advisory committee was appointed by the U.S. Environmental Protection Agency (EPA) to assess the effects of the insecticide dichlorodiphenyltrichloroethane (DDT). The committee used a food chain model of the bioaccumulation of DDT developed by Oak Ridge National Laboratory (O'Neill and Burke 1971) to assist in deciding whether to immediately ban the insecticide or end its use gradually (Holcomb Research Institute 1976).

1.2.8 SYSTEM IDENTIFICATION

In many systems, there will be a component for which the input and output can be observed, but the internal structure or mechanisms are not understood. Such a component is referred to as a *black box*. By conducting simulation experiments with different component structures, and using model validation methods, the most likely internal structure may be identified. According to May (2004, p. 791), "Various conjectures about underlying mechanisms can be made explicit in mathematical terms, and the consequences can be explored and tested against the observed patterns. In this general way, we can, in effect, explore possible worlds."

1.2.9 UNANTICIPATED EFFECTS

A really useful reason for modeling and simulation would be if we could predict unexpected events. Systems, however, can be quite complex with many interconnections among subsystem components. Therefore, it is often difficult to anticipate how a perturbation in one subsystem will affect other subsystem components. This is particularly true of unusual or catastrophic events. These occur so infrequently in real systems, it makes it difficult to observe and measure their impacts. And yet, simulation experiments of these extreme events, with model parameters set at extreme values, may be most important. First, these simulations could determine the level of perturbation that could cause irreversible damage to the system. And second, investigating system behavior at these thresholds or breakpoints can lead to greater understanding of system behavior.

The purpose for which a model is developed will greatly affect the rigor of the validation required of the model. We discuss the implications for model validation further in Chapter 10.

1.3 WHAT DOES IT TAKE TO BE A MODELER?

A modeler in environmental toxicology must have a strong foundation in three areas: first, you must have a strong foundation in the relevant biological toxicological processes; you also must have the requisite skills in math and statistics; and third, you must have the necessary skills in computer programming. Readers of this text are expected to have a good foundation in the basic toxicological sciences and some training in math and statistics. In this text we will cover the math and statistics techniques that are important in modeling, and give the reader a good foundation in programming, using the MATLAB® programming language.

This text is not intended for other modelers but for toxicologists and decision makers who wish to learn something about modeling. The reader also should be able to evaluate existing models and critically assess simulation results. This one text therefore will not in itself make one a modeler. What it may do is give the ecological toxicologist a basic understanding of what is involved in modeling in environmental toxicology. It also will make it easier for him or her to communicate with modelers. And, it may even inspire some toxicologists to develop a modeling knowledge base by completing the requisite math, statistics, and computer programming courses. For an extensive discussion on training in biomathematics, see Lucas (1962). If a toxicologist does not possess adequate knowledge in these disciplines, he or she will have to rely on others, who may not have a toxicology background, to participate in the modeling process.

The problem of communication between toxicologist and mathematician or statistician also has a long history and stems primarily from a different way of thinking about and developing models (van der Vaart 1977). Toxicologists conduct experiments on a process that yield data that can then be used to infer the structure of a model. This involves inductive reasoning. Mathematicians, on the other hand, derive their models from first principles and logic using deductive reasoning. There are advantages and disadvantages to each way of thinking about models. A modeler in environmental toxicology should be cognizant of the different patterns of thought, especially when trying to communicate with mathematicians and statisticians.

What about a modeler who is part of a modeling effort in environmental toxicology? Obviously, it would be better to have the modeler be knowledgeable in environmental toxicology to reduce the possibility of errors in communication between toxicologist and modeler. Likewise, it is better if a decision maker has some understanding of what is involved in developing models used in decision making.

The third foundation needed is the ability to write computer programs in some language that is amenable to modeling and simulation. A more detailed discussion of programming languages is included in Chapter 2, with an overview of the languages used in this text, MATLAB and Simulink.

1.4 WHY MODELS FAIL: A CAUTIONARY NOTE

Most cases of model failure involve models that fail to accurately predict future events. Models designed for other purposes do not receive much attention when they fail, mostly because the potential consequences of using a faulty model are much less than when models fail at making accurate predictions. In other cases, the failure to consider model predictions has resulted in bridges collapsing, dams and levees bursting, and toxic waste ponds leaking (Petroski 1992, Pilkey and Pilkey-Jarvis 2007).

1.4.1 Poor Data for Parameter Estimation

One reason models fail is the use of poor data for parameter estimation. The expression "garbage-in, garbage-out" is more than a catchy phrase. It accurately describes the results of using inappropriate data for estimating system initial conditions or model parameters. Data can come from a number of sources, including laboratory or field studies; these can be either historical data or data from experiments designed specifically for parameter estimation. Data from laboratory studies usually have less variability than those from field studies but they may not reflect real system variables. Field studies, on the other hand, yield data with greater variability and are usually more costly than lab studies. Historical data, from published literature, have unique problems. They should be from lab or field studies that are as close to the modeled system as possible. In some cases, even these data may not be available and parameters will have to be estimated by extrapolating from other systems. In these cases, a conservative approach should be taken in which parameter values are estimated that can provide simulation results that reflect a worst-case scenario.

1.4.2 Uncertainty Not Considered

One of the greatest causes of model failure is the use of constant parameter values when the parameter is a random variable. A real-world variable with high variability should be represented in a model by parameters with a comparable level of variability (see Chapter 4), not just its mean value. Models with variable parameters can produce simulations significantly different from models with constant values for those same parameters. That is because at the tail ends of a parameter's probability distribution, the dynamical behavior of the system can be quite different from that using just the mean of the distribution. It is at the tails of a distribution that parameters can cause the most extreme perturbations of a system.

1.4.3 Bias (Political, Social, Economic)

Some models are developed with a particular point of view and the accuracy of these models' predictions is rarely tested. The models are developed, either consciously or subconsciously, to obtain the desired results. The simulation results are then used to justify the point of view. Political bias can come into play when state and federal regulatory agencies differ from those being regulated over the best way to control pollution. Different models can be developed to show different outcomes of regulatory policies. Similarly, models are used to make predictions about the outcomes of social policies. Models can be developed to overestimate the consequences of not following a given policy, which often can be used to justify increases in the budgets of those making policy. Perhaps the most egregious source of bias is when a model is used to support a position that shows a product is harmless when, in fact, it may cause harm to human health or the environment. A product judged to be harmless can mean significant economic benefits to the company marketing that product.

1.4.4 Lack of Understanding of Real-World Systems

The failure of predictive models to accurately predict the future results more from a lack of understanding of the system being modeled than from inaccuracies in computer programming. Assuming that we have an error-free computer program, it will predict only what it has been designed to do, that is, the model can make predictions about a system based only upon the assumptions in the model. That is why it is important for a modeler, or a model user, to have a good understanding of the system being modeled—to assure that the appropriate mechanisms are built into the model so that it is capable of addressing the pertinent questions posed by the modeler. In the case of modeling wild populations or ecosystems, it is imperative that the modeler spend time in the natural system to observe its components and their relationships.

1.4.5 Misuse of Mathematics

The population geneticist, Charles Birch, and ecologist, Henry Andrewartha, warned ecologists to "avoid the *misuse* of mathematics. … Ecologists, and especially mathematicians with a slight knowledge of biology, seem to be prone to the mistake of building a model with symbols which, they pretend, represent certain qualities of animals" (Andrewartha and Birch 1954, 11). Fifty years later, however, Bialek and Botstein (2004, p. 789) argued, "Understanding how to reason in the language of mathematical symbols is essential, but one must go further to appreciate that these symbols actually stand for variables of the natural world." The different perspectives reflect the increased complexity and realism of today's models compared with those in 1954. Increased complexity and computing power, as well as available simulation software, however, do not guarantee more reliable predictions. Instead of just applying "elegant" mathematics, a modeler must have a clear understanding of what is going on in the real world (May 2004).

REFERENCES

Andrewartha, H. G., and L. C. Birch. 1954. *The Distribution and Abundance of Animals*. Chicago: University of Chicago Press.

Bialek, W., and D. Botstein. 2004. "Introductory Science and Mathematics Education for 21st-Century Biologists." *Science* 303:788–790.

Chapman, D. G. 1971. "Mathematics and Ecology." In *Statistical Ecology, Vol. 3*, ed. G. P. Patil et al., 428–434. University Park, PA: Pennsylvania State University Press.

Dixon, K. R., and B. D. Murphy. 1979. "A Discrete-Event Approach to Predicting the Effects of Atmospheric Pollutants on Wildlife Populations Using ^{14}C Exposure." In *Animals as Monitors of Environmental Pollutants*, 15–26. Washington, DC: National Academy of Sciences.

Epstein, J. M. 2008. "Why model?" *Journal of Artificial Societies and Social Simulation* 11(4):12, http://jasss.soc.surrey.ac.uk/11/4/12.html.

Gentner, D., and A. Stevens, eds. 1983. *Mental Models*. Hillsdale, NJ: Lawrence Erlbaum Assoc.

Hermann, C. 1967. "Validation Problems in Games and Simulation with Special Reference to Models of International Politics." *Behavioral Science* 12:216–230.

Holcomb Research Institute. 1976. *Environmental Modeling and Decision Making*. New York: Praeger.

Killough, G. G., K. R. Dixon, N. T. Edwards, et al. 1976. *Progress Report on Evaluation of Potential Impact of 14C Releases from an HTGR Reprocessing Facility*. ORNL/TM-5284, Oak Ridge, TN: Oak Ridge National Laboratory.

Lucas, H. L., ed. 1962. *The Cullowhee Conference on Training in Biomathematics*. Raleigh, NC: Institute of Statistics of North Carolina State College.

Martin, F. F. 1968. *Computer Modeling and Simulation*. New York: Wiley.

May, R. M. 2004. "Uses and Abuses of Mathematics in Biology." *Science* 303:790–793.

Morgan, M. G., B. Fischoff, A. Bostrom, et al. 2002. *Risk Communication: A Mental Models Approach*. Cambridge: Cambridge University Press.

O'Neill, R. V., and O. W. Burke. 1971. *A Simple Systems Model for DDT and DDE Movement in the Human Food Chain*. ORNL-IBP-71-9, Oak Ridge, TN: Oak Ridge National Laboratory.

Petroski, H. 1992. *To Engineer Is Human: The Role of Failure in Successful Design*. New York: Vintage Books.

Pilkey, O. H., and L. Pilkey-Jarvis. 2007. *Useless Arithmetic: Why Environmental Scientists Can't Predict the Future*. New York: Columbia University Press.

Rieber, L. P., S.-C. Tzeng, and K. Tribble. 2004. "Discovery Learning, Representation, and Explanation within a Computer-Based Simulation: Finding the Right Mix." *Learning and Instruction* 14:307–323.

van der Vaart, H. R. 1977. "Some Signposts for the Education of Systems Ecologists." In *New Directions in the Analysis of Ecological Systems, Part 1*, ed. G. S. Innis, 35–41. La Jolla, CA: Simulation Council.

Windschitl, M., and T. Andre. 1998. "Using Computer Simulations to Enhance Conceptual Change: The Roles of Constructivist Instruction and Student Epistemological Beliefs." *Journal of Research in Science Teaching* 35:145–160.

2 Principles of Modeling and Simulation

2.1 SYSTEMS

This chapter describes principles of modeling and simulation. It is important, however, to understand that these activities are grounded in the concept of systems. Therefore, we begin by describing what a system is and why knowledge of a system is important.

2.1.1 DEFINITION

The term *system* has become commonplace in many scientific disciplines. In physiology there are cardiovascular, respiratory, and immune systems as well as the central nervous system. In ecology, there are ecosystems. In meteorology, there are weather systems. There are expert systems and geographic information systems. What do all of these systems have in common? Each system is a collection of interconnected components, or subsystems, that functions as a complete entity or whole that is greater than the sum of its separate parts. The cardiovascular system consists of the heart, lungs, blood, and connecting veins and arteries. An ecosystem is a collection of plant and animal populations that interact through various processes such as predation, consumption, competition, and decomposition. A formal definition of a system will aid in later discussions of modeling and simulation.

Definition 2.1: A *system* is a collection of components that are interconnected in such a way as to function as a whole. ∎

The adverse effects of many toxicants are directly related to their ability to interfere with the normal functioning of systems. The receptor–ligand interactions of the nervous system, for example, are affected by organophosphates. Many toxic chemicals interfere with cellular energy production systems. Other toxic substances can impair the reproductive system by reducing spermatogenesis or causing birth defects or abortion of the fetus.

2.1.2 SYSTEM INPUT AND OUTPUT

The components or compartments of a system usually are represented by state variables, that is, those variables that define the state of the system. Once a system has been defined, it is possible to identify stimuli or disturbances, called *inputs*, from outside the system, that operate on the system to produce a response called the *output*. For example, the nematode fumigant dibromochloropropane (DBCP) can be an input to an animal, through ingestion or inhalation, with a resulting output of decreased spermatogenesis.

2.1.3 CONTROL SYSTEMS

One special type of system is one in which one or more inputs to the system are controlled or regulated by the system to produce one or more desired outputs. In engineering, this type of system is called a *control system*. A typical example of a control system is a thermostat-controlled heating system. The temperature is regulated or controlled by the thermostat, which provides the input reference temperature. The input operates on the heating system, which produces the output—ambient temperature. Control systems also are found at all levels of biological organization. At the individual organism level, endocrine systems are control systems. For example, the hypothalamus produces a gonadotropin-releasing hormone, which is input to the anterior pituitary gland, regulating the production of gonadotropins follicle stimulating hormone (FSH) and luteinizing hormone (LH), which in turn regulate the production of gonadal hormones. At the ecosystem level, a predator stalking or attacking its prey is another example of a control system. The input is the observed position of the prey. This visual stimulus operates on the predator's central nervous system and muscles to produce the output of movement of the predator toward the prey. These examples are illustrated with block diagrams later in this chapter.

2.1.4 Feedback

That component of the control system responsible for activating the system to produce the output is the *control action*. In the previous examples, the control action in the heating system is the thermostat, in the hypothalamus-pituitary-gonadal system it's the androgen and estrogen receptors in both the hypothalamus and pituitary, and in the predator-prey system it's the predator's brain. A system in which the control action is dependent upon the output is called a *closed-loop* or *feedback control system*.

Definition 2.2: *Feedback* in a closed-loop system is the signal obtained by comparing the output with the input.   ■

Feedback can be either positive or negative. Negative feedback tends to increase stability of the system, whereas positive feedback decreases stability. The systems described previously are all negative feedback systems. In the heating system, the feedback signal is the output temperature, which is compared with the reference temperature in the thermostat. In the hypothalamus-pituitary-gonadal system, the feedback signal is the actual level of gonadal hormone, which is compared with the normal level at the pituitary. The feedback signal in the predator-prey system is the position of the prey, which feeds back to the predator's brain.

2.1.5 System States: Steady State versus Transient States

In modeling and simulation of systems, we are interested in analyzing and forecasting a system's dynamics. The behavior of a system over time can tell us the effects of a toxicant on that system. It can be difficult to attribute observed dynamic behavior to a disturbance from a toxicant, however, because of inherent fluctuations or transients in the system. A system in which the initial transients have disappeared and no new disturbances are input will be in a condition where the system components do not change with respect to time. Such a system is said to be in a *steady state*. We will look at transient behavior of a system in more detail in Section 2.2.1.2.

2.1.6 Discrete versus Continuous

Components of a system can fluctuate in a continuous fashion such as water flow in a river system. Other components may increase or decrease at discrete intervals such as a population of fish that spawns over a short period of the year in the river system.

In most complex systems, there will be both continuous and discrete components. In addition to changes with respect to time, systems also can fluctuate in space. An animal population that is growing in size is likely to expand into new habitats. Or, a population may migrate and change its geographic location. This is an important consideration in simulating contaminant effects because exposure can change as an animal changes locations. It follows that a system must be described as continuous or discrete in both time and space.

2.1.7 LINEAR VERSUS NONLINEAR

Another way of classifying systems is by their response to inputs, whether the inputs are controlled variables, random variables, or disturbances. Those systems that satisfy the principle of superposition are termed *linear systems*. This principle states that the response of a linear system resulting from several simultaneous inputs is equal to the sum of the responses to each of the individual inputs. Those systems that do not satisfy the principle of superposition are termed *nonlinear systems*. All real systems are nonlinear to some degree because for extreme input values the principle of superposition will not hold. Some systems may be approximately linear under normal environmental conditions. This distinction is important in ecotoxicology because our main concern is with those situations that are not normal. Therefore, it is important to determine whether relations between system components are nonlinear because nonlinear *threshold* responses may occur. Also, some toxic chemicals may be antagonistic or synergistic. We will discuss linear systems in some detail because an understanding of the behavior of linear systems provides a conceptual basis for the study of nonlinear systems.

2.2 MODELING

Because the dynamics of real systems are quite complex, understanding the impacts of toxicants on a system can be enhanced by modeling the system. In applying modeling to ecotoxicology, we are interested in studying a "real-world" system and the effects of various toxicants on that system. The modeling approach we take in this book is to define the system relationships in terms of quantitative mechanisms in a model of the system. A model is a necessary abstraction of the real system. For a model to adequately reflect the behavior of a system perturbed by a toxicant, however, requires a mechanistic approach to modeling. The level of abstraction is determined by the objectives of the model. A model designed to give a quick sketch of the dominant effects on a system will require less realism than one designed to provide accurate predictions of future system behavior.

The modeling process involves three steps: (1) identification of system components and boundaries, (2) identification of component interactions, and (3) characterization of those interactions using quantitative abstractions of mechanistic processes. Once the model has been defined, it is implemented on a computer. This is the process of simulation (see Section 2.3).

Models can be described in a variety of ways. In this book, we use two approaches: equations and block diagrams.

2.2.1 EQUATIONS

Equations can define the quantitative functional relations among system components as well as represent the system dynamics, or changes in mass or energy of the system components as functions of time. Difference equations are used to describe discrete systems whereas differential equations are used to represent continuous systems. For some applications, where the numerical value of the state variables remains high, difference equations can approximate continuous systems.

The behavior of models described by the two types of equations depends upon the time step used in the model, and the difference is usually small for small time steps. This can be seen by a study of the definition of differential equations.

Definition 2.3: A *differential equation* is an algebraic equality (equation) that contains either differentials or derivatives. ∎

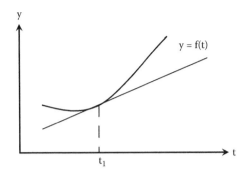

FIGURE 2.1   The function $y = f(t)$ showing the derivative of y, which is the slope at the point t_1.

Definition 2.4: The *derivative* of a function f is another function, the differential operator d/dt, whose value at any time t_1 in the domain of f is:

$$\frac{df(t)}{dt} = \lim_{0 \leftarrow \Delta t} \frac{f(t + \Delta t) - f(t)}{\Delta t} \qquad (2.1) \blacksquare$$

In other words, the derivative of the function $f(t)$ is the function at some time, t plus Δt, minus the function at time t, divided by Δt, and then taking the limit of that function as Δt approaches zero, provided this limit exists.

Let $y = f(t)$. The derivative dy/dt can be interpreted as the instantaneous rate of change of y with respect to the independent variable t. The derivative of the function $f(t)$ also can be defined for t_1 as the slope of the tangent to the curve $f(t)$ at t_1 (Figure 2.1).

A first-order system (one with a single compartment) can be represented by the general first-order differential equation:

$$\frac{dy(t)}{dt} = f(y(t), t) \qquad (2.2)$$

As pointed out previously, the derivative dy/dt is equivalent to the instantaneous rate of change of the function $f(y(t), t)$. The following example of a differential equation describes the decrease in material from a single compartment, or the exponential decay process.

Example 2.1

Here we have an example of the *exponential decay function* (also called the *exponential elimination function*). The exponential decay function can be written as

$$y(t) = y_0 e^{-rt} \qquad (2.3)$$

where y_o is the initial value of y at time $t = 0$ and r is the decay rate constant. The derivative of y is then defined as the negative rate constant times y:

$$\frac{dy}{dt} = -r \cdot y \qquad (2.4)$$

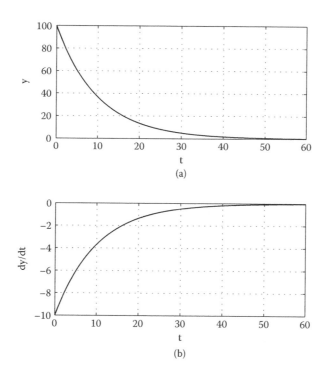

FIGURE 2.2 (a) Exponential decay function, *y*, and (b) its derivative, *dy/dx*.

Figure 2.2 illustrates the relation between a function and its derivative. Note that the derivative (rate) takes on the value of zero when the function is parallel to the *t* axis (zero slope).

There are several ways of classifying differential equations. One way is based on the number of independent variables. A partial differential equation is one with more than one independent variable. A differential equation with only one independent variable is called an *ordinary differential equation*. Each type can have one or more dependent variables. Examples of independent variables are the Cartesian axes in three-dimensional space and time. Partial derivatives can be used to model the movement of toxicants in geographical space but are beyond the scope of this text.

A second type of classification of differential equations depends on the way the independent variable time is treated. If time is expressed explicitly in the differential equation, it is a *time-variable differential equation*. A *time-invariant differential equation* is one that has no terms that explicitly include time as an independent variable. In other words, time is only expressed in the derivative of the dependent variable (e.g., *dy/dt* where *t* is time).

In the last section we defined linear and nonlinear systems. There is a close correspondence between systems and differential equations. Linear differential equations can represent linear systems and nonlinear differential equations can represent nonlinear systems. Just as linear systems can approximate nonlinear systems under certain conditions, linear differential equations sometimes can approximate nonlinear systems. Even at the point where nonlinearities are included in a model, a comparison with the dynamics of a linear model can be instructive. A linear differential equation is one that has no terms that include higher powers, products, or transcendental functions of the dependent variables. Any differential equation that has any such term is a nonlinear differential equation.

In general, an *n*th-order linear system can be described by a linear differential equation of order *n*:

$$\frac{d^n y}{dt^n} + a_1 \frac{d^{n-1} y}{dt^{n-1}} + \cdots + a_{n-1} \frac{dy}{dt} + a_n y = f(t) \tag{2.5}$$

where *f(t)* is a linear input function, with constant coefficients. Equations of an order greater than one are rarely used in ecological modeling (but see Clark 1971, Innis 1972). If, in Equation 2.5, there is zero input (i.e., $f(t) = 0$), the equation is said to be *homogeneous*. In this text, we place the emphasis on nonhomogeneous, first-order differential equations.

In general, a first-order linear system (i.e., a system with 1 compartment) can be described by a linear differential equation of order 1:

$$\frac{dy}{dt} + ay = f(t) \tag{2.6}$$

where, in general, a is a function.

Although the coefficients in Equation (2.6) are, in general, functions, preliminary models often begin by assuming a is a constant. It follows therefore, that ordinary, linear, constant coefficient, differential equations are important in the modeling of dynamic systems.

2.2.1.1   Solution of Ordinary First-Order Differential Equations

There are two ways of solving differential equations: analytical solutions and numerical solutions. Analytical solutions are possible for relatively simple differential equations used to model the relationships between variables. Some examples of analytical solutions are described in Chapter 5. A system of nonlinear differential equations generally is too complex to be solved analytically. Numerical integration methods are required for these models and are discussed in greater detail in Section 2.2.1.4. We can also use the analytical solution to linear, constant coefficient, ordinary differential equations as an approximation to nonlinear differential equations with variable coefficients. The analysis of linear, constant coefficient, differential equations also will provide us with the foundation for the later analysis of more complex systems. In this section, we describe the solution, $y = y(t)$, to the nonhomogeneous, linear, constant coefficient, ordinary differential equation (Equation [2.6]) where $f(t)$ is a linear input function, also with constant coefficients. The general solution $y(t)$ to Equation 2.6 consists of two parts, the free response $y_a(t)$ and the forced response $y_b(t)$. The free response is the solution to Equation (2.6) when there is zero input and the solution depends only on the initial condition, $y(0)$. The forced response $y_b(t)$ of the differential Equation (2.6) is the solution of the differential equation when all the initial conditions are zero. Whereas the free response has zero input, the forced response is defined by the input $f(t)$. The sum of these two responses comprises the total response or solution of the equation.

Example 2.2

Again, let's look at the example of the exponential elimination function. This time we use the letter Q for the dependent variable and include the parameter p_4, which is the minimum or final value of Q. The initial value of Q is p_5. For our example of an exponential elimination or one-compartment elimination model, we define the rate of change in the state variable Q with the differential equation:

$$\frac{dQ}{dt} = -p_2(Q - p_4) \tag{2.7}$$

where
p_2 = the elimination rate constant, and p_4 = the minimum or steady-state value of Q.

Now, the free response, $Q_a(t)$ is:

$$Q_a(t) = p_5 e^{-p_2 t} \tag{2.8}$$

where
p_5 = the initial value of Q, and the forced response, $Q_b(t)$ is

$$Q_b(t) = p_4 \left[1 - e^{-p_2 t} \right] \tag{2.9}$$

The total response is the sum of the free response and the forced response:

$$Q(t) = Q_a(t) + Q_b(t)$$
$$= p_5 e^{-p_2 t} + p_4[1 - e^{-p_2 t}]$$
$$= p_5 e^{-p_2 t} + p_4 - p_4 e^{-p_2 t}$$
$$= (p_5 - p_4)e^{-p_2 t} + p_4$$

(2.10)

We can now plot each part of the solution (Figures 2.3 and 2.4).

1. Initial conditions: $Q(0) = 100$ and $Q(0) = 50$. Input, $p_2 = 0.05$, $p_4 = 10$.
   Note that starting with two different initial conditions, the free response (a) approaches zero for both curves, showing that the response only depends upon the initial condition. The forced response (b) does not change with the different initial conditions but starts at zero and approaches the value of the parameter p_4. The total response is the sum of the free response and the forced response where the total response shows the different initial conditions but approaches the p_4 value of 10 rather than zero.
2. Initial condition: $Q(0) = 100$. Input, $p_2 = 0.05$, $p_4 = 10$ and $p_4 = 20$.
   In this case, the input, p_4, has no effect on the free response, which again approaches zero. The forced responses approach the two values of p_4. The total response for both input values starts at the initial condition, $Q(0) = 100$, but each response approaches a different input value of p_4 (Figure 2.4).

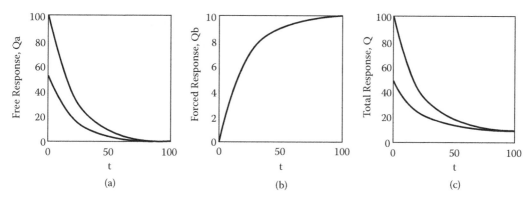

FIGURE 2.3 Solution to elimination or decay function: (a) free response, (b) forced response, and (c) total response.

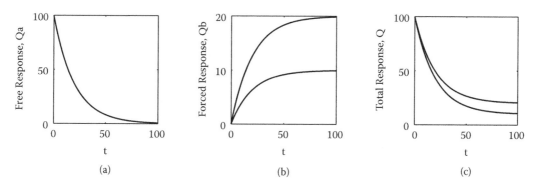

FIGURE 2.4 Solution to elimination or decay function: (a) free response, (b) forced response, (c) total response.

2.2.1.2 Steady-State and Transient Response

The total response also can be described by the sum of two other responses: the transient response and the steady-state response. The *transient response* is that part of the total response that approaches zero as time approaches infinity. The *steady-state response* is that part of the total response that does not approach zero as time approaches infinity.

The total response in Example 2.2 was

$$Q(t) = (p_5 - p_4)e^{-p_2 t} + p_4 \qquad (2.11)$$

The transient response, then, is

$$Q_T(t) = (p_5 - p_4)e^{-p_2 t} \qquad (2.12)$$

which approaches zero because as time increases, the exponential function approaches zero. The steady-state response is the constant p_4, which does not approach zero:

$$Q_{ss} = p_4 \qquad (2.13)$$

The total response again is the sum of the transient response and the steady-state response.

2.2.1.3 Difference Equation Approximation to Differential Equation

A differential equation can be approximated by a difference equation, especially for small values of Δt. The approximation can be seen by giving Δt a value of one in Equation 2.1 and then not taking the limit:

$$\frac{dy(t)}{dt} = \lim_{0 \leftarrow \Delta t} \frac{y(t + \Delta t) - y(t)}{\Delta t}$$

$$\frac{dy(t)}{dt} \approx \frac{y(t+1) - y(t)}{1} \qquad (2.14)$$

$$\frac{dy(t)}{dt} \approx y(t+1) - y(t)$$

By substituting the discrete approximation $y(t + 1) - y(t)$ for the derivative dy/dt, Equation (2.6) now can be approximated by the difference equation:

$$\frac{dy(t)}{dt} \approx y(t+1) - y(t) = ay$$

$$y(t+1) = y(t) + ay \qquad (2.15)$$

Example 2.3: Exponential Decay Function

To show the close approximation of a difference equation to a differential equation, we plotted the solution to the differential Equation (2.7) with the difference equation approximation (Figure 2.5).

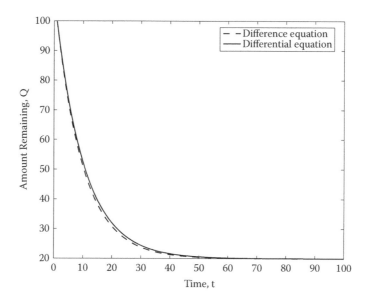

FIGURE 2.5 Exponential decay function plotted as the numerical solution to a differential equation and as the difference equation approximation.

2.2.1.4 Numerical Solutions to Differential Equations

In Section 2.2.1.1, we discussed the analytical solution to ordinary differential equations. In simulation, differential equations can be difficult to solve analytically. These equations may require numerical solutions, which are usually close approximations to analytical solutions. In the chapters that follow, we use numerical methods exclusively to solve the differential equations in a model. There are many different numerical methods used to obtain solutions, and many variations on each method. Conceptually, numerical methods start from an initial solution of the differential equation to estimate the dependent variable $y(t_n)$ at time point n and then take a short step forward in time to find the next solution $y(t_{n+1})$ at time point $(n + 1)$. We briefly describe the algorithms most commonly used in this text.

Numerical methods can be grouped according to whether the equations are stiff or nonstiff. A *stiff* equation is a differential equation whose solution can become unstable if there are drastic changes in the rate of change (slope) of the variables or if the algorithm step size is extremely small. For *nonstiff* differential equations, variations on the Runge-Kutta methods are important iterative methods. These techniques were developed by the German mathematicians Carl David Tolmé Runge and Martin Wilhelm Kutta. The basic algorithm divides the time interval between predicted values of the dependent variable $y(t)$ into smaller intervals for which the slope is estimated. If there are four intervals in the time step, the method is called a fourth-order method. The method is referred to as a one-step solver because to estimate $y(t_n)$, it needs only the solution at the immediately preceding time point, $y(t_{n-1})$. Other numerical methods, such as the Adams-Bashforth-Moulton predictor-corrector solver, are multistep solvers in that they normally need the solutions at several preceding time points to compute the current solution. For some equations, the nonstiff methods might result in a magnification of errors in approximation, leading to an unstable or nonexistent solution. Methods that do not magnify approximation errors are called *numerically stable*. It is important to use a stable method when solving a stiff equation. If the nonstiff methods do not work, or are extremely slow, one should try an implicit method designed for stiff equations such as the backward Euler method or implicit Runge-Kutta method. These methods are described further in Chapter 3.

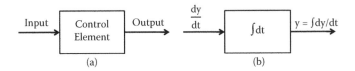

FIGURE 2.6   Block diagram elements: (a) single block with an input and an output, and (b) a block with a integral operator. The integral operates on the input derivative to create an output *y*. (From Joseph J. DiStefano III et al., *Feedback and Control Systems 2nd ed.* Schaum's Outline Series. New York: McGraw-Hill, 1990. With permission from author.)

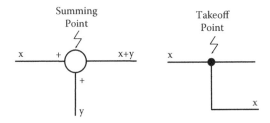

FIGURE 2.7   Block diagram summing point and takeoff point. (Modified from Joseph J. DiStefano III, et al., *Feedback and Control Systems 2nd ed.* Schaum's Outline Series. New York: McGraw-Hill, 1990. With permission from author.)

2.2.2   BLOCK DIAGRAMS

A second method of representing a model of a system is a flowchart in which blocks representing system components are connected by lines representing the flow of information between components. A *block diagram* is a particular type of flowchart that illustrates the functional relations among system components, particularly where relationships show cause and effect. Blocks can represent state variables, other system components, and mathematical operations (Figure 2.6). As we will show, there can be a direct transformation from a differential (or difference) equation to a block diagram and vice versa.

Additional features of a block diagram are summing points and takeoff points (Figure 2.7). A *summing point* is where two or more inputs are summed to produce a single output. A *takeoff point* is where a single input is branched to provide identical inputs to two or more blocks or summing points.

A block diagram of a feedback system has a takeoff point from the output signal that feeds back to a summing point where it is summed with the reference input (Figure 2.8). A negative feedback system has a negative feedback signal and a positive feedback signal represents a positive feedback system.

The general structure of a block diagram is shown in Figure 2.8. This general structure includes all of the basic elements of a block diagram, although not all elements are necessarily found in all systems. Also, most systems will be much more complex with several parallel paths representing many state variables and their interactions.

Controlled system. Each subsystem can be represented by a controlled system or controlled process. Inputs to the controlled system include an internal control signal and external disturbances. Output from the controlled system usually is the state variable used to describe the subsystem.

Control elements. Control elements are the subsystem components that generate the control signal input to the controlled system.

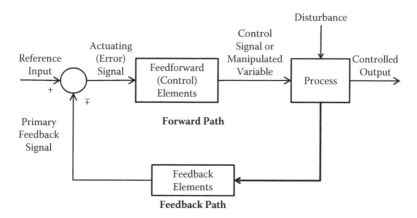

FIGURE 2.8 Generalized block diagram. (From Joseph J. DiStefano III, et al., *Feedback and Control Systems 2nd ed.* Schaum's Outline Series, New York: McGraw-Hill, 1990. With permission from author.)

Control signal. The control signal is produced by control elements and acts upon the controlled process.

Feedback elements. The feedback elements are the subsystem components that define the functional relationship between the controlled output and the primary feedback signal.

Controlled output. The controlled output is usually the state variable used to define the subsystem.

Primary feedback signal. The primary feedback signal results from the action of the feedback elements operating on the controlled output. The signal is compared with the reference input at the summing point.

Reference input. The reference input is an external stimulus to the subsystem. The reference input is compared with the primary feedback signal at the summing point.

Actuating signal. This signal is the input from the summing point to the control elements and results from the summation of the reference input and the primary feedback signal.

Disturbance. Disturbance inputs are external environmental variables or toxic perturbations. These inputs can affect the controlled system directly or indirectly through the effects on other system components.

In Sections 2.1.3 and 2.1.4, we described three examples of negative feedback control systems, a thermostat, the hypothalamus-pituitary-thyroid system, and a predator behavior system. The following examples illustrate these systems using block diagrams

Example 2.4: Thermostat

A block diagram for a thermostat is shown in Figure 2.9 (DiStefano et al. 1990).

In the thermostat model, the controlled output is the actual room temperature. The reference input is the temperature that is set to the desired, or reference, temperature. The actual room temperature then becomes a feedback signal in a negative feedback loop. The reference temperature and the actual temperature are compared in a summing point (thermostat) to generate an actuating signal. If the set temperature is higher than the actual temperature, the actuating signal will be positive so the control element (furnace) is actuated (turned on). The control signal, or manipulated variable, is the heat from the furnace, which raises the room temperature. When the actual room temperature exceeds the set temperature, the actuating signal now is negative and the furnace turns off.

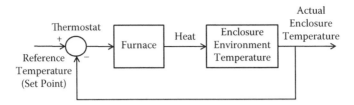

FIGURE 2.9   Block diagram for a thermostat. (From Joseph J. DiStefano III, et al., *Feedback and Control Systems 2nd ed.* Schaum's Outline Series. New York: McGraw-Hill, 1990. With permission from author.)

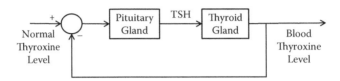

FIGURE 2.10   Block diagram for the hypothalamus-pituitary-thyroid control system. (From Joseph J. DiStefano III, et al., *Feedback and Control Systems 2nd ed.* Schaum's Outline Series. New York: McGraw-Hill, 1990. With permission from author.)

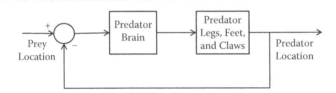

FIGURE 2.11   Block diagram for a predator behavior control system. (Modified from Joseph J. DiStefano III, et al., *Feedback and Control Systems 2nd ed.* Schaum's Outline Series, New York: McGraw-Hill, 1990. With permission from author.)

EXAMPLE 2.5: Hypothalamus-Pituitary-Thyroid System

A simplified diagram of the hypothalamus-pituitary-thyroid system is shown in Figure 2.10 (DiStefano et al. 1990). A more complete description of this system can be found in Carr and Norris (2006).

In this simplified system, the controlled output is the blood thyroxine level. It is transported in the bloodstream to the brain in a negative feedback loop. There it is compared with the normal thyroxine level (reference input). If the actuating signal is positive, the blood thyroxine level is greater than normal. This signals the pituitary gland to reduce the secretion of the thyroid stimulating hormone (TSH). This control signal reduces the activity of the thyroid, lowering the amount of thyroxine secreted into the bloodstream.

EXAMPLE 2.6: Predator-Prey System

Block diagrams can also be used to model animal behavior. In this example, we model the behavior of a predator as it hunts its prey (Figure 2.11).

In this system, the controlled output is the location of the predator. The reference input is the location of the prey. The relative distance between the two locations is an actuating signal processed by the predator's brain. The predator's brain sends a control signal to the predator's legs, feet, and claws to move the predator closer to the prey. The distance is continually monitored and the predator adjusts his position until he is able to attack the prey.

2.2.3 STOCHASTIC MODELS

Deterministic modeling involves the use of parameters that take on only a single value, thus "determining" the model's outcome. Stochastic modeling, on the other hand, involves the use of random variables. Not only can model coefficients be functions of other variables, they can be functions of random variables and thus can be random variables themselves.

Definition 2.5: A *random variable* is a variable that takes on values with some relative frequency. ■

In this way of classifying models, those with random (stochastic) variables are called *stochastic models* and those without are called *deterministic models*. Random variables are used to represent the random variation or "unexplained" variation in the state variables. We describe stochastic models more fully in Chapter 4.

2.2.4 INDIVIDUAL-BASED MODELS

Models that simulate all individuals simultaneously are referred to as *individual-based models*. Each individual in the simulation has a unique set of characteristics: age, size, condition, social status, and location in the landscape. Each individual has its own history of daily foraging, reproduction, and eventual mortality.

A number of individual-based models have been described by Huston and DeAngelis (1988) and DeAngelis and Gross (1992). Individual-based models have been applied to ecotoxicology by Hallam et al. (1989) and Hallam and Lassiter (1994), among others. This approach is becoming popular for several reasons. For one thing, it enables the modeler to include complex behavior and decision making by individual organisms. Most importantly, it allows one to model populations in complex landscapes, where different individuals may be experiencing very different conditions. Individual-based models were very uncommon up until a few years ago because they require a great amount of computer power. As computers increase in power, however, these models are becoming prevalent. A flowchart for a generic individual-based model is shown in Figure 2.12. A specific model of this type, which is described in Chapter 11, has been developed to study avian populations exposed to agricultural insecticides.

2.2.5 AGGREGATED MODELS

There are two general ways in which models of individuals can be extended to a population as a whole. First, one can simulate not just one individual, but all individuals that make up the population of interest. Second, one can aggregate various population members into classes, such as age classes. The model then follows not individual organisms but variables representing the numbers of individuals per age class. An example of an aggregated population model is described in Chapter 13.

2.3 SIMULATION

Once we have a model defined in quantitative terms, the next step in using the model in the study of systems is simulation. We use simulation to obtain the system output as a function of time, or the time response of the system. We do this by exercising the model on a computer.

There are many things to consider before a simulation can be performed. These include determining the type of computer and programming language, cost of running the computer, the data needed for model parameters, and availability of expertise for the design and analysis of simulation experiments.

The types of computers available for simulation are analog, digital, and hybrids of analog and digital computers. Digital computers are the most widely used in simulation, including applications

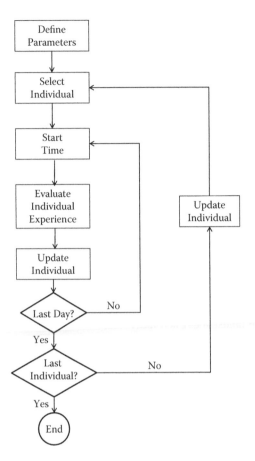

FIGURE 2.12 Flowchart for generic individual-based model. (From Jeffrey A. Tyler and Kenneth A. Rose, Individual Variability and Spatial Heterogeneity in Fish Population Models. *Reviews in Fish Biology and Fisheries* 4 (1994):91–123. Reprinted with permission from Springer.)

in ecotoxicology, although there are some limitations that must be considered. One problem is the sequential nature of digital computation. In large complex systems, with many processes operating simultaneously, it is difficult to determine the most logical sequence for computation. The recent advance in parallel processing computers has resolved this problem to a great extent. Although these computers are available at major universities via networks, access is still rather limited. Another consideration is the numerical integration of differential equations. This is not a serious limitation but does require making decisions about integration method and time interval.

Computer programming languages can be divided into three classes: (1) programming languages, (2) simulation languages, and (3) simulation packages. Programming languages include FORTRAN, BASIC, PASCAL, C, and C++. These languages require more expertise in programming than simulation languages, but in general, are more efficient and require less time for execution. Examples of system simulation programming languages include ACSL, CSMP, DYNAMO, GASP, SIMAN, and SLAM. Most simulation packages have been developed for specific applications such as industrial plant operations, aeronautical and space simulation, and electronic component simulation. A number of these packages are general enough for use in a wide range of applications. Two simulation packages that have been used in ecotoxicological applications are STELLA® and RAMAS®. STELLA (isee systems, Lebanon, New Hampshire) primarily is used in education rather than research. It describes models using links among system compartments rather than equations, although equations can be generated from the compartment diagram. RAMAS

Ecotoxicology (Applied Biomathematics, Inc., Setauket, New York) models age-structured populations primarily for ecological risk assessments. The simulation packages featured in this text, MATLAB® and Simulink® are described further in Chapter 3.

2.3.1 Principles of Simulation

2.3.1.1 Principle of Communication

We previously described several purposes for the simulation of systems. In each case, there was an implicit assumption that the simulation results would be used by someone—researchers, planners, managers, politicians, lawyers, and so on. For a model and the simulation results to be accepted by the end user, they must have confidence in the model's validity and have some general understanding of how the model was developed and implemented. It is important, therefore, that the decision maker is involved in all steps of the simulation process and communication between modeler and end user is maintained throughout.

2.3.1.2 Principle of Modularity

Each system can be divided into a number of subsystems. Each subsystem must be defined in the same way as the whole system by identifying the subsystem components and boundary. Depending upon the complexity of the system, subsystems could be divided into sub-subsystems and so on. This modularity lends itself to improved model development and implementation. By concentrating on a small part of the system at a time, it is less likely that we will fail to identify significant variables and their interrelations. If we develop models of subsystems as relatively independent entities, it will be easier to validate each submodel than the entire system model. It also will be much easier to write and debug the computer program code using object-oriented programming techniques.

2.3.1.3 A Modified Principle of Parsimony

How does one decide which variables to include in a model? We have stated that an important first step in the modeling process is the identification of the state variables and boundaries of the system being studied. Unfortunately, there are no definite rules for their identification. For the sake of simplicity, we would like to include as few variables as possible and still capture the basic dynamics of the system. This is where modeling becomes more of an art than a science. Obviously, the modeler or another researcher should be familiar with the system being modeled. Those most familiar with a system, however, may overlook some important variable because they are too close to the problem. Someone unfamiliar with the system, with a different perspective on the problem, should not be hesitant to suggest the possibility of interrelations with new variables and of including them in the model. A balance must be struck between making the model unnecessarily complex by adding more variables and possibly leaving out some significant variable. The likelihood of leaving out an important variable is reduced when, in the development of the conceptual model, a mental image of the system is created. In this mental image, the flow of the toxicant, if traced throughout the system, looks for all possible physical and biological pathways.

2.3.2 Steps in Simulation

Each simulation problem is unique. Therefore, the steps involved in each simulation problem may vary, depending upon the objectives of the problem. There are, however, several basic steps that comprise the simulation process in most problems. These include problem definition, model development, model implementation, determination of data requirements, estimation of parameters, model validation, design of simulation experiments, analysis of results, and presentation and implementation of simulation results (Figure 2.6). We describe these steps briefly in the rest of this section and in greater detail in later chapters.

2.3.2.1 Problem Definition

The step of defining the problem begins with a statement of the objective of the simulation study. This statement should be clear and as precise as possible, although the goal may be revised or additional goals may be added as the study progresses. The system also must be defined, including subsystems (see Section 2.1).

2.3.2.2 Model Development

In this step, we evaluate the modeling approach that best addresses the problem (or even whether a modeling approach is appropriate). A conceptual model is then described in abstract terms in which the relevant state variables and controlled and uncontrolled input variables must be identified as well as the hypothesized interactions among the variables. An important, but often neglected, step is the development of a logical flow diagram that shows the arrangement of the subsystems and the flow of materials among them.

2.3.2.3 Model Implementation

In model implementation, an explicit model is formulated by developing a quantitative expression for the system followed by quantitative descriptions of subsystem mechanisms and the interrelationships among subsystem components. These expressions are then combined into equations or block diagrams defining the dynamics of the subsystems. The subsystems then are combined into a simulation model of the whole system. The subsystem mechanisms can usually be modeled using one of the equations in Table 2.1.

The equations in the first column can be used as terms in a dynamic process. The independent variable, x, in this case is time, which appears explicitly in the equation. It is logical that a time-independent equation in which time does not appear explicitly will have the advantage of being able to model a process that is not tied directly to time. Therefore, the derivative form of the model (column 3) is used because the independent variable does not appear explicitly in most cases. As we

TABLE 2.1
Commonly Used Functions and Their Derivatives, along with Their Graphs

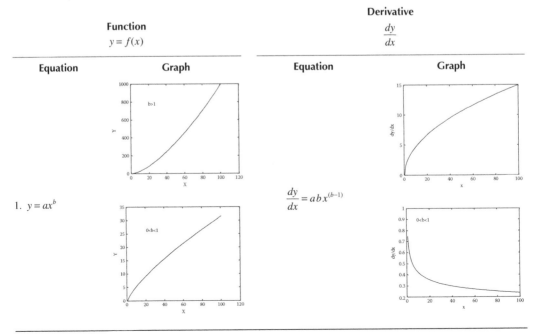

TABLE 2.1 *(Continued)*
Commonly Used Functions and Their Derivatives, along with Their Graphs

Function $y = f(x)$		Derivative $\dfrac{dy}{dx}$	
Equation	Graph	Equation	Graph
2. $y = y_0 e^{ax}$		$\dfrac{dy}{dx} = ay$	
3. $y = y_0 e^{-ax}$		$\dfrac{dy}{dx} = -ay$	
4. $y = y_{max}(1 - e^{-bx})$		$\dfrac{dy}{dx} = b(y_{max} - y)$	
5. $y = \dfrac{y_{max}}{1 + e^{a-bx}}$		$\dfrac{dy}{dx} = by(y_{max} - y)$	
6. $y = ax^b e^{cx^d}$		$\dfrac{dy}{dx} = \dfrac{y}{x}\left[cdx^d + bx^b\right]$	

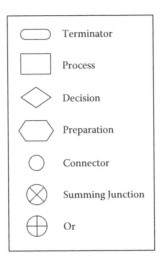

FIGURE 2.13   Some common flowchart symbols.

will see in Chapter 6, parameters often are easier to estimate in the integral form in column 1 than in the derivative form in column 3. In practice then, the parameters are estimated using the integral form and then are used in the derivative form in the model.

At this step in the simulation process, a test of model validity involves examination of the conceptual model, the logical flowchart, and the model structure for logical errors or inconsistencies. A check of variable and parameter units for consistency is important, particularly if parameter values are obtained from different sources that may not have reported the parameter values in the same units. Additional steps in model implementation include: programming the computer code for the model, verification of the computer code, and documentation. Steps in development of the computer code include selection of a simulation language, construction and verification of a program flowchart, and writing the program code. A flowchart can range from a "gross" to a detailed level depending upon the complexity of the model and the experience of the programmer. Even a flowchart for a complex model may start at the "gross" level and detail added later as development progresses. An example of a flowchart is Figure 2.11 for a general individual-based model. An experienced programmer usually can benefit from a flowchart by avoiding programming errors or costly program "debugging." Flowchart symbols are not standardized, although some have become de facto standards through common use (Figure 2.13).

Verification of the computer code involves a comparison of the logical and program flowcharts as well as program code tests intended to challenge the program code and other parts of the overall system functionally and structurally. Functional testing demonstrates only that the system outputs appear to be correct. It does not allow an assessment of whether the program code is actually performing according to specifications and requirements. A complete functional test of every combination of inputs may not be feasible except for very small programs. Functional testing is essentially a subset of structural testing.

Structural testing is designed to exercise all submodules and branches of the program code and their interrelationships with the hardware and peripheral devices. Structural testing is performed to ensure that all relevant functions in the simulation program perform as intended. Each of the following tests should be conducted to prove that the system is working properly.

1. *Normal testing* includes cases that test the functional and structural integrity of the simulation system. The input data for these test cases all fall within the range that the user considers normal. Performing enough test cases can give a reasonable level of confidence that the system behaves as intended under normal conditions.

2. *Boundary testing* is performed using values that force the system to discern whether the input is valid or invalid, or to make a decision as to which branch of the program to execute. Boundary test values are set at the edges (i.e., slightly below and above) of valid input ranges. Boundary testing does not mean making the simulation system "crash" or involuntarily stop.

3. *Special case testing*, also known as *exceptional case testing*, documents the system's reactions to specific types of data or lack of data and is intended to ensure that the simulation system does not accept unsuitable data. These tests should be designed to document what happens when values that are not included in the ranges defined in the specifications are entered. Use of test cases with no data entry in a field will assist in establishing simulation system defaults.

4. *Parallel testing* is one of the most common types of tests performed by computer program developers. Parallel testing is performed by running two systems in parallel and comparing the outputs (e.g., two simulation program versions or simulation output compared with a manual procedure). The comparison of the outputs from the same simulation program on different computer systems or different versions of the program on the same system is part of parallel testing plans.

Documentation at this step should include the following:

1. A logical flowchart with a description of validity checks
2. Equations or block diagrams
3. A program flowchart with a description of validity checks
4. Complete program code with internal documentation
5. A list of system variables and parameters with units
6. Instructions for running the program

In Chapters 11, 12, and 13, we will go through three examples of model implementation.

2.3.2.4   Data Requirements

Once the system variables and parameters have been identified and explicit mathematical expressions written, values for them must be obtained. Initial values for all state variables must be measured and constants and parameters must be estimated. In addition, an independent set of values for the state variables needs to be collected for comparison with simulation outputs in the validation process. Sources of historical data for the system being studied (or a similar system) must be identified and experiments designed to obtain missing data. The decision maker or other user of the simulation output must be consulted for the types of exogenous inputs or perturbations to be included in the system simulation. Parameter estimation is discussed further in Chapter 7.

2.3.2.5   Model Validation

Volumes have been written about what constitutes *model validation*. Generally, validation is a process in which measurements obtained from the real system are compared to the model projections. Several iterations of comparing model behavior to that of the real system are usually required to obtain a satisfactory or *valid* model. For example, if after the first iteration, major differences are found between the data and model output, the model is modified in a way that the modeler believes will improve the fit of the model to the data. Simulation experiments are run and the new model output is compared to the data. This process continues until the model is judged to be valid. The process of model validation does not occur at a single point in the simulation process. In fact, validation occurs at nearly every step in the simulation process. We will explore model validation in more detail in Chapter 10.

2.3.2.6   Design of Simulation Experiments

Experimental design in simulation is very similar to other applications of classical statistics. With all of the possible combinations of variables and parameter values, a screening process is needed to identify the optimal combination to study in a simulation experiment. The objective in designing a simulation experiment is to gain knowledge about the relationship between input disturbance from toxic substances and the output values of the state variables that describe the system. In this step in the process the model is tested for validity. The output is examined to identify errors and possibly revise the experimental design or the model structure. We will examine the design of simulation experiments further in Chapter 8.

2.3.2.7   Analyze Results of Simulation Experiments

The analysis of simulation experiments can tell us a lot about the system being simulated. In addition to the statistical analysis of the simulation output, there are several other types of postsimulation analysis. We would like to know something about the stability of the system. How far from equilibrium does the system move in response to a toxicant disturbance and how long does it take for the system to return to equilibrium? How sensitive is the model (and the real system) to variations in system parameters? Is there a point where the system moves to a different equilibrium and is it possible for the system to return to the original equilibrium? Other types of analysis include the determination of the optimal number of simulation runs and reduction of the variability associated with system output. The analysis of simulation experiments is described more fully in Chapter 9.

2.3.2.8   Presentation and Implementation of Results

If the results of our simulations and the recommendations upon which they are based are to be implemented, it is important that the simulation results be presented in a way that they are easily understood by the decision maker. This allows the decision maker to communicate any questions he or she may have about the simulation and how that may be interpreted for the real system. This dialogue is essential if the next iteration in the updating of the model is to incorporate those changes that will bring about the desired results. Simulation is an area where the adage of a picture being worth a thousand words holds true. Simulation results should be presented in graphical form, particularly to illustrate system dynamics in both space and time. Some simulation languages include graphical output. In fact, this feature is one of the advantages of using simulation software. In the case of spatial modeling, models can be linked to geographic information systems to display simulation output in a map format. In systems where toxicants are tracked in three dimensions, such as lakes, soils, and the atmosphere, three-dimensional digital imaging and visualization methods can be used.

REFERENCES

Carr, J. A., and D. O. Norris. 2006. "The Hypothalamus-Pituitary Axis." In *Endocrine Disruption: Biological Bases for Health Effects in Wildlife and Humans*, ed. D. O. Norris and J. A. Carr, 58–86. New York: Oxford University Press.

Clark, J. P. 1971. "The Second Derivative and Population Modeling." *Ecology* 52:606–613.

DeAngelis, D. L. 1992. *Dynamics of Nutrient Cycling and Food Webs*. New York: Chapman & Hall.

DeAngelis, D. L., and L. J. Gross. 1992. *Individual-Based Models and Approaches in Ecology: Populations, Communities, and Ecosystems*. New York: Chapman & Hall.

DiStefano III, J. J., A. R. Stubberud, and I. J. Williams. 1990. *Feedback and Control Systems*, 2nd ed., Schaum's Outline Series. New York: McGraw-Hill.

Hallam, T. G., and R. R. Lassiter. 1994. "Individual-Based Mathematical Modeling Approaches in Ecotoxicology: A Promising Direction for Aquatic Population and Community Ecological Risk Assessment." In *Wildlife Toxicology and Population Modeling: Integrated Studies of Agroecosystems*, ed. R. J. Kendall and T. E. Lacher, pp. 531–542. Boca Raton, FL: Lewis Publishing.

Hallam T. G., R. R. Lassiter, and S. A. L. M. Kooijman. 1989. "Effects of Toxicants on Aquatic Populations." In *Applied Mathematical Ecology*, ed. S. A. Levin, T. G. Hallam, and L. J. Gross, 352–382. Berlin: Springer-Verlag.

Huston, M. D., and D. L. DeAngelis. 1988. "New Computer-Models Unify Ecological Theory: Computer-Simulations Show That Many Ecological Patterns Can Be Explained by Interactions among Individual Organisms." *BioScience* 38:682–691.

Innis, G. 1972. "The Second Derivative and Population Modeling: Another View." *Ecology* 53:720–723.

Tyler, J. A., and K. A. Rose. 1994. "Individual Variability and Spatial Heterogeneity in Fish Population Models." *Reviews in Fish Biology and Fisheries* 4:91–123.

3 Introduction to MATLAB® and Simulink®

In this chapter, we provide an overview of MATLAB® and Simulink®. MATLAB and Simulink are registered trademarks of The MathWorks, Inc. This chapter is not intended as a complete user's guide, but to orient the reader to their basic structure. The manuals provided with the products give excellent guidance: MATLAB *Getting Started Guide* (The MathWorks 2010a) and *Simulink Getting Started Guide* (The MathWorks 2010b). In addition, The MathWorks, Inc. Web site offers comprehensive documentation, demos, and tutorials. Some of the most widely used topics in this text are introduced in this chapter and developed further in later chapters.

3.1 MATLAB

MATLAB is technical computing software that provides high-level modeling and simulation capabilities that are relatively easy to learn compared to C, C++, and FORTRAN. It is particularly good in visualizing data and simulation output using 2-D and 3-D graphics functions. The version of MATLAB used in this text is Version 7.10.0.499 (R2010a). It was run on the Microsoft Windows XP Version 5.1 operating system.

MATLAB is used in a wide range of applications that can be extended using the several add-on toolboxes that are available, including the Bioinformatics Toolbox™, the Statistics Toolbox™, the Neural Network Toolbox™, and the Curve Fitting Toolbox™. The only toolbox required for this text is the Statistics Toolbox. We used Version 7.3 (R2010a) of the Statistics Toolbox in this text. The SimBiology® product is a systems biology and toxicokinetics programming tool that does modeling and simulation primarily using block diagrams. It provides much of the modeling and simulation functionality covered in this text. It allows for stochastic modeling and supports parameter estimation, sensitivity analysis, parameter scans, and other model analysis methods. It is recommended that one use SimBiology only after learning the basic modeling tools covered in this text.

MATLAB is initiated either from a desktop shortcut or by selecting MATLAB in the list of installed programs using Start. When opened, there are four windows in view: the Command Window, a Current Folder browser, a Workspace browser, and a Command History (Figure 3.1).

The Command Window is used to type commands such as MATLAB functions and to display the results of commands or statements in a script file. Commands can be repeated by pressing the up arrow key until the command appears in the Command Window; the command is then executed by pressing the Enter key. If a set of commands is likely to be repeated, a program containing all the commands can be written in a script file called an *m-file* using the Editor (Figure 3.2). In the Editor, you can open an existing program by clicking **File** and then **Open**, or by clicking the **Open File** icon. You can create a new file by clicking **File** and then **New** or by clicking the **New File** icon. To run the m-file, click the **Run** icon. By running the m-file, all the commands in the file are executed in sequence.

The Current Folder browser lists all the files in the Current Folder. To change the current folder, double-click that folder. The individual files can be opened by double-clicking the file name in the list of files in the current folder. The current folder also can be selected using the Current Folder field in the MATLAB toolbar. The Workspace browser lists the parameters used in the current session with their current value. It also gives the dimensions of the variables and their minimum and

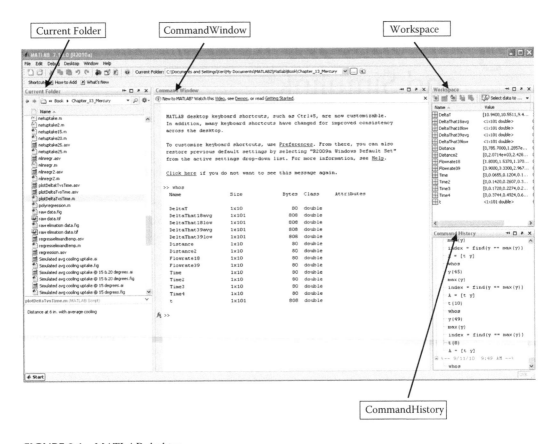

FIGURE 3.1   MATLAB desktop.

maximum values. You can clear the workspace by typing the command **clear** in the Command Window. The Command History lists recent commands typed in the Command Window. The commands are listed by date and the time that a particular session started.

3.1.1 Matrix Algebra

Most MATLAB operations utilize *matrices*, defined as a rectangular array of numbers. This generally makes functions more efficient, but does require thinking in terms of matrix algebra. A *matrix* is defined in terms of the number of rows and columns. As an example, the following matrix has four rows and four columns:

$$
A = \begin{bmatrix}
5 & 3 & 11 & 8 \\
2 & 16 & 4 & 7 \\
12 & 9 & 10 & 1 \\
14 & 13 & 6 & 15
\end{bmatrix}
$$

A matrix with a single row or single column is called a *vector*. In matrix A, the first row can be written as a 1×4 vector, that is, a vector with one row with four elements: $A(1) = [5 \quad 3 \quad 11 \quad 8]$. Matrix A can be created in MATLAB by reproducing the above matrix, either in the Command Window or an m-file:

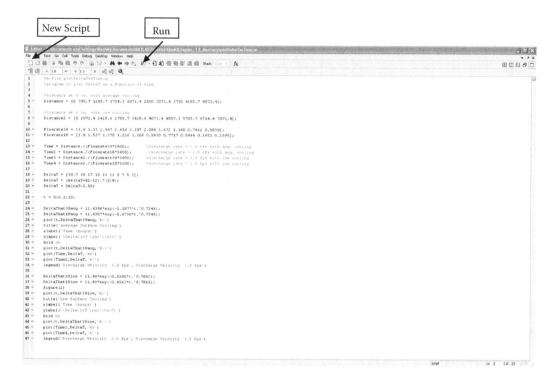

FIGURE 3.2 MATLAB Script File Editor.

A =

5	3	11	8
2	16	4	7
12	9	10	1
14	13	6	15

A second way of creating the matrix is to write *A* as a vector with rows separated by semicolons: A = [5 3 11 8; 2 16 4 7; 12 9 10 1; 14 13 6 15].

One can select certain elements of the matrix using the colon operator (:). For example, to select the elements of the first row, columns 2 through 4, type **y = A(1,2:4)** in the Command Window. The answer is:

y =

3	11	8

Because MATLAB is matrix based, matrix multiplication and division involve two different types of arithmetic operations. Matrix arithmetic operations are defined by the rules of linear algebra. Array arithmetic operations are carried out element by element. The period character (.) distinguishes the array operations from the matrix operations. To multiply two matrices, A and B, the syntax is **A*B** where the number of columns of A must equal the number of rows of B. Array multiplication is the element-by-element product of the arrays A and B. The syntax is **A.*B** where A and B must be the same size. Similarly, array right division, **A./B** is the matrix with elements **A(i,j)/B(i,j)**. A and B must have the same size. Matrix division, **A/B**, is equivalent to the array B postmultiplied by the inverse of **A, B*inv(A)**.

For example, given the matrix A above and the matrix B:

$$B = \begin{bmatrix} 2 & 6 & 1 & 4 \\ 3 & 5 & 10 & 6 \\ 7 & 3 & 8 & 2 \\ 1 & 9 & 6 & 5 \end{bmatrix}$$

```
A*B  =  [104     150     171     100
          87     167     236     147
         122     156     188     127
         124     302     282     221]
```

and

```
A.*B  =  [10      18      11      32
           6      80      40      42
          84      27      80       2
          14     117      36      75]
```

```
A/B  =  [0.7877      1.8153     -0.1231     -1.1592
         0.8153     -0.7707      0.1529      1.6115
        -0.1295     -1.4427      2.2049      1.1529
         4.7919      1.3790      0.4193     -2.6561]
```

and

```
A./B  =  [2.5000      0.5000     11.0000      2.0000
          0.6667      3.2000      0.4000      1.1667
          1.7143      3.0000      1.2500      0.5000
         14.0000      1.4444      1.0000      3.0000]
```

3.1.2 DATA INPUT

In many of the techniques needed to develop models and run simulations, we use data, particularly in parameter estimation and model validation. These data have to be accessed by the programs written to implement these techniques. There are several different ways of inputting data in MATLAB. One way is to use the Import Wizard to input data to the Workspace. Once data files have been created, they can be located by choosing **Select File > Import Data**. If the data are to be used in a script file, they can be written as part of the file or loaded into the workspace using the **load** function. The following **load** command will load data from a MAT-file into the workspace: **load(filename,variables)**. To load a file in ASCII format, use the following command: **load(filename,'-ascii')**. ASCII files must contain a rectangular table of numbers, with an equal number of elements in each row. The file delimiter (character between each element in a row) can be a blank, comma, semicolon, or tab. For example, the following statements load a text file mercurydata.txt with two columns representing the variable *day* in column one and *concentration* in column two. The variables then are plotted using the **plot** command.

```
load ('mercurydata.txt')
day = mercurydata(:,1);
conc = mercurydata(:,2);
plot(day,conc,'o')
```

If the text file is not in the current directory, you can change the directory or specify the file with the complete location string.

Another important input data format can be Excel spreadsheets. To input the data into a matrix, use the **xlsread** command as follows to read the data in the Excel file, called "file"

```
ndata=xlsread(file)
```

For example, the command

```
ndata = xlsread('mercurydata.xls')
```

reads the same data as the example above, but from the spreadsheet mercurydata.xls. Again, if the text file is not in the current directory, you can change the directory or specify the file with the complete location string. The **xlsread** statement ignores the column heading. The headings can be input as well using the command

```
[ndata,text]=xlsread('mercurydata.xls')
```

where **text** contains the headings. The results are displayed in the Command Window:

```
ndata =
        25          0          1       9699
        25          0          2      11060
        25          0          3       9458
        25          1          1       4616
        25          1          2       5281
        25          1          3       5624
        25          2          1       4170
        25          2          2       5169
        25          2          3       5566
        25          3          1       2332
        25          3          2       2868
        25          3          3       4199
        25          7          1       1567
        25          7          2       1570
        25          7          3       2299
        25          9          1       1313
        25          9          2        827
        25          9          3       1014
        25         11          1        557
        25         11          2        529
        25         11          3        531
        25         14          1        588
        25         14          2        648
        25         14          3        436
        25         16          1        911
        25         16          2       1099
        25         16          3        302

text =

     'Temperature'     'Day'     'Replicate'     'Hg concentration'
```

If there is more than a single sheet in the spreadsheet, the above statement will automatically input the first sheet. To get data from another sheet in the spreadsheet, called "uptake," located in the directory c:\MATLAB\mercurydata.xls, the statement

```
[ndata,text]=xlsread('c:\MATLAB\mercurydata.xls','uptake')
```

will generate the following output in the Command Window:

```
ndata =

  1.0e+003 *

    0.0250    0.0015    0.0010    1.0590
    0.0250    0.0015    0.0020    1.0840
    0.0250    0.0015    0.0030    1.2010
    0.0250    0.0030    0.0010    1.8590
    0.0250    0.0030    0.0020    1.7790
    0.0250    0.0030    0.0030    2.0210
    0.0250    0.0045    0.0010    2.3390
    0.0250    0.0045    0.0020    2.4900
    0.0250    0.0045    0.0030    2.4520
    0.0250    0.0060    0.0010    2.9180
    0.0250    0.0060    0.0020    3.1830
    0.0250    0.0060    0.0030    2.5110
    0.0250    0.0120    0.0010    3.7500
    0.0250    0.0120    0.0020    4.7770
    0.0250    0.0120    0.0030    3.8260
    0.0250    0.0180    0.0010    5.0570
    0.0250    0.0180    0.0020    4.9930
    0.0250    0.0180    0.0030    5.4200
    0.0250    0.0240    0.0010    5.8970
    0.0250    0.0240    0.0020    5.3340
    0.0250    0.0240    0.0030    6.0950
    0.0250    0.0320    0.0010    5.3160
    0.0250    0.0320    0.0020    5.3160
    0.0250    0.0320    0.0030    5.8860
    0.0250    0.0480    0.0010    5.5530
    0.0250    0.0480    0.0020    5.5240
    0.0250    0.0480    0.0030    6.4230
    0.0250    0.0720    0.0010    7.8590
    0.0250    0.0720    0.0020    7.3880
    0.0250    0.0720    0.0030    8.0750

text =

    'Temperature'    'Hour'    'Replicate'    'Hg concentration'
```

If you only need a specified region in the sheet (such as a subset of the variables), specify the rows and columns with the statement:

```
[ndata,text]=xlsread('c:\MATLAB\mercurydata.xls','elimination',
'B2:D28')
```

which generates the following output:

```
ndata =

            0            1         9699
            0            2        11060
            0            3         9458
            1            1         4616
            1            2         5281
            1            3         5624
            2            1         4170
            2            2         5169
            2            3         5566
            3            1         2332
            3            2         2868
            3            3         4199
            7            1         1567
            7            2         1570
            7            3         2299
            9            1         1313
            9            2          827
            9            3         1014
           11            1          557
           11            2          529
           11            3          531
           14            1          588
           14            2          648
           14            3          436
           16            1          911
           16            2         1099
           16            3          302

text =

    {}
```

The output is the same as that generated in the first example above except that the variable in column 1 (Temperature) was not output. Note also that since we did not include the header row in the statement, no text was output.

3.1.3 SOLVING EQUATIONS

In Section 2.2.1.4, we introduced numerical solutions to differential equations. In this section we describe some of the MATLAB differential equation solvers. The solver used most often in this text is fourth-order Runge-Kutta, **ode45**, used mostly to solve nonstiff differential equations. The statement:

```
[tout,yout] = ode45(odefun,tspan,y0)
```

solves the system of equations

$$\frac{dy}{dt} = f(t,y)$$

in the function **odefun**.

Each row in the solution array **yout** corresponds to a time returned in the column vector **tout**. The solution is integrated over the time span, **tspan**, defined by the vector **[t0 tfinal]** where **t0** is the starting time for the simulation and **tfinal** is the ending time. The initial conditions for the state variables **y** are defined by the vector **y0**.

For example, the function **resp3** describes a differential equation model of inhalation uptake (Example 5.1).

```
%resp3
function Up=resp3(~,U)
Y=3.0;          %Exposure concentration, mg/m^3)
Vt=22;          %Tidal volume, ml/breath
f=8.;           %Breathing frequency, breaths/minute
c=.01;          %Elimination rate constant, 1/min
Up=(1e-6)*Y*Vt*f - c*U;
```

The solver **ode45** is used to find the solution to the differential equation, and the results are plotted using the **plot** statement:

```
%resp4
U0=0;
tspan=[0 100];
[t,U]=ode45(@resp3,tspan,U0);
plot(t,U);
xlabel ('Time');
ylabel('Lung Concentration');
grid;
```

The output vector U contains the integrated values of dU/dt at the time points generated by the solver **ode45**. The time points will not be equally spaced, but are close together where the slope changes relatively quickly and are farther apart where the slope changes more slowly. By typing **[t,U]** in the Command Window, the time points and the U values are displayed as a two-column matrix. In this case there were 49 time points. There may be cases, however, when the y values are needed at specific times, such as having different output generated at the same time points. To obtain solutions at specific times $t_0, t_1, \ldots, t_n$, the time points in **tspan** can be specified using the colon operator {:}, **tspan=(0.0:10:100.0)**. The output in this case is generated at 10-minute intervals from 0 to 100.

Taking the above example, the solver statements are:

```
%resp4_2
U0=0;
tspan=(0:10:100);
[t,U]=ode45(@resp3,tspan,U0);
plot(t,U);
xlabel ('Time');
ylabel('Lung Concentration');
grid;
```

Typing [t,U] in the Command Window now shows that the 11 time points and the 11 U values were output.

As we pointed out in Chapter 2, there are many different algorithms available to solve differential equations. The following list describes the algorithms used in MATLAB. For nonstiff problems, the MATLAB solvers are **ode45, ode23,** and **ode113**.

ode45 is based on an explicit Runge-Kutta (4,5) formula. It is a *one-step* solver in that it needs only the solution at the immediately preceding time point, $y(t_{n-1})$ in computing $y(t_n)$. In general, **ode45** is the best function to apply as a *first try* for most problems.

ode23 is another implementation of an explicit Runge-Kutta (2,3) algorithm. It may be more efficient than **ode45** at crude tolerances and in the presence of moderate stiffness. Like **ode45, ode23** is a one-step solver.

ode113 is a variable-order Adams-Bashforth-Moulton predictor-corrector (PECE) solver. It may be more efficient than **ode45** at stringent tolerances and when the ODE file function is particularly expensive to evaluate. **ode113** is a *multistep* solver; it normally needs the solutions at several preceding time points to compute the current solution.

If the nonstiff solvers appear to be unduly slow, try using one of the stiff solvers **ode15s, ode23s, ode23t,** or **ode23tb**.

ode15s is a variable-order solver based on the numerical differentiation formulas (NDFs). Optionally, it uses the backward differentiation formulas (BDFs, also known as Gear's method) that are usually less efficient. Like **ode113, ode15s** is a multistep solver. Try **ode15s** when **ode45** fails, or is very inefficient and you suspect that the problem is stiff.

ode23s is based on a modified Rosenbrock formula of order two. Because it is a one-step solver, it may be more efficient than **ode15s** at crude tolerances. It can solve some stiff problems for which **ode15s** is not effective.

ode23t is an implementation of the trapezoidal rule using a "free" interpolant. Use this solver if the problem is only moderately stiff and you need a solution without numerical damping.

ode23tb is an implementation of TR-BDF2, an implicit Runge-Kutta formula with a first stage that is a trapezoidal rule step and a second stage that is a backward differentiation formula of order two. By construction, the same iteration matrix is used in evaluating both stages. Like **ode23s**, this solver may be more efficient than **ode15s** at crude tolerances.

In deciding which solver to use, it is recommended to start with **ode45**. If that solver doesn't work, try any of the others until you find one that does work. If none of them work, you will need to change the tolerances or modify the differential equation. All differential equations in this text are solved with either **ode45** or **ode23**.

3.1.4 SAVING DATA

After generating simulation output, one might wish to save the output to a file for postprocessing analysis. The MATLAB command for saving files is appropriately named **save**.

The single command, **save,** saves the variables in the workspace in a MATLAB-formatted binary file (MAT-file) called **matlab.mat**. To specify a file name, the command **save('filename')** stores all variables from the current workspace in a MAT-file called **filename**. The file will be saved in the current folder; it can be saved in another location by specifying the full path in the **filename**. One can specify variables that are to be saved in a file with the command **save(filename,variables)**, which stores only the specified variables.

For example, load the file **ndata** using the command

```
[ndata,text]=xlsread('c:\MATLAB\mercurydata.xls','elimination')
```

and define the four variables: Temperature, Day, Replicate, Hg concentration

```
Temp=ndata(:,1)
Day=ndata(:,2)
Rep=ndata(:,3)
Conc=ndata(:,4)
```

Now save the file **ndata** keeping only the variables **Day** and **Conc**:

```
save('ndata','Day','Conc')
```

The file **ndata** now contains only **Day** and **Conc**. You can confirm this by first clearing the workspace by typing **clear all** in the Command Window, and then loading **ndata** into the workspace by typing **load ndata** in the Command Window. Now if you type **whos** in the Command Window, only the variables **Day** and **Conc** appear.

If you want to add more data or variables to an existing file, use the command **save('filename','variables','-append',)**. Instead of the default .mat format, you can save a file in ASCII format. The command

```
save('filename','variables','-append','-ascii' )
```

appends the specified variables in filename with **'-ascii'** format.

3.1.5 Plotting Data

MATLAB offers an extensive set of plotting options. If one is interested in plotting two variables, x and y, the **plot(x,y)** command will create a two-dimensional plot of vector **y** versus vector **x**. Both vectors must be the same length (i.e., have the same number of elements). The **plot(y)** command plots the columns of **y** versus their index. If there are three variables, x, y, and z, the **plot3(x,y,z)** command will create a three-dimensional plot. All three vectors must be of the same length. The plot can be edited using the plot editor or plotting options can be written into the m-file script. If you are going to run the same plotting routine several times, it may be more efficient to write the plot changes into the m-file script. One way to initiate plot editing is to click the **Edit Plot** icon in the figure window and then select **View > Property Editor** on the **Figures** toolbar (Figure 3.3). The second way of initiating plot editing is to type **plottools** in the Command Window when a figure window is open. At this point the figure can be edited to change line style, font size, add text, and so on.

Programmable syntax can vary the properties of a plotted line, including line style, line color, line width, plot symbol, and nearly any other property of a line. The options for the line properties are shown in Table 3.1.

For example, the following statements will create a plot with a red dotted line connecting circles at each set of x,y values. The line width is of 2 points, and the circles are of size 10 and have black edges and green faces.

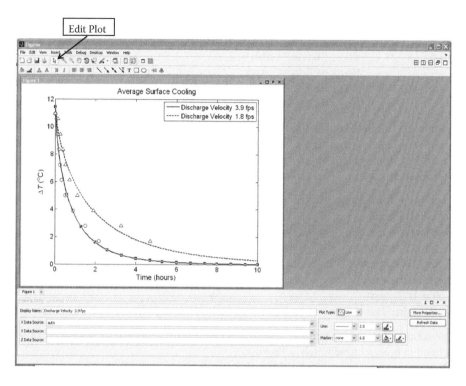

FIGURE 3.3   The Plot Editor opened to edit Figure 1 using the Property Editor.

TABLE 3.1
Colors, Plot Symbols, and Line Types for Plotting Lines

Color Symbol	Color	Marker Symbol	Marker	Line Symbol	Line Type
b	blue	o	circle	-	solid
g	green	x	x-mark	:	dotted
r	red	+	plus	-.	dashdot
c	cyan	*	star	--	dashed
m	magenta	s	square	(none)	no line
y	yellow	d	diamond		
k	black	v	triangle (down)		
w	white	^	triangle (up)		
		<	triangle (left)		
		>	triangle (right)		
		p	pentagram		
		h	hexagram		

```
x = 0:0.5:12;
y = 1-(0.5*exp(-0.5*x).*sin(0.5*x+0.5));
plot(x,y,'--ro','LineWidth',2,...
            'MarkerEdgeColor','k',...
            'MarkerFaceColor','g',...
            'MarkerSize',10)
```

In addition to plotting lines for simulation output and data, there is a wide range of options for creating text in figures, such as figure titles, axis labels, figure legends, and annotation text. These properties include font type, font color, subscripts and superscripts, and special characters. Some font properties are:

- **\FontName{fontname}:** This property is the name of the font, such as Courier. The default font is Helvetica.
- **FontAngle:** There are two main font angles: normal and italic.
  \rm: This property is the normal font.
  \it: This property is the italic font.
- **\FontSize{fontsize}:** This property is the size in font units. The default point size is 10 (1 point = 1/72 inch).
- **FontWeight:** The weight can be light, {normal}, demi, or bold.

The subscript character (_) and the superscript character (^) modify the character or substring that is defined in braces immediately following.

The default text interpreter is TeX, although it can be set to the LaT$_E$X interpreter. To print the special characters used to define the TeX strings, prefix them with the backslash (\) character.

Table 3.2 shows some of the most commonly used symbols, including lowercase and uppercase Greek letters.

Some text examples from Chapter 13 will illustrate some of the syntax previously described. To write the label ΔT °C, use the syntax \Delta\itT \rm(^{\o}C). To write μg·kg^{-1}, use the syntax, \mug\cdotkg^{−1}.

TABLE 3.2
Some Special Symbols Used in Modifying Text

Symbol	Character Sequence	Symbol	Character Sequence	Symbol	Character Sequence
α	\alpha	σ	\sigma	Ψ	\Psi
β	\beta	τ	\tau	Ω	\Omega
γ	\gamma	υ	\upsilon	≈	\approx
δ	\delta	φ	\phi	~	\sim
ε	\epsilon	χ	\chi	≤	\leq
ζ	\zeta	ψ	\psi	∞	\infty
η	\eta	ω	\omega	°	\circ
θ	\theta	Γ	\Gamma	±	\pm
ι	\iota	Δ	\Delta	≥	\geq
κ	\kappa	Θ	\Theta	∂	\partial
λ	\lambda	Λ	\Lambda	·	\bullet
μ	\mu	Ξ	\Xi	÷	\div
ν	\nu	Π	\Pi	≠	\neq
ξ	\xi	Σ	\Sigma	o	\o
π	\pi	Υ	\Upsilon	∫	\int
ρ	\rho	Φ	\Phi	©	\copyright

FIGURE 3.4   Simulink Library Browser window.

3.2   SIMULINK

In Chapter 2, we described the block diagram method of model development. The MATLAB software that solves these block diagram models is Simulink. We used Simulink Version 7.5 (R2010a) to construct and run the models in this text. Simulink is started by typing **simulink** in the MATLAB Command Window. The Simulink Library Browser will appear in a window showing different classes of blocks (Figure 3.4).

Block diagrams are constructed using a graphical user interface (GUI) to solve difference or differential equations. The interactive graphical environment simplifies the modeling process, eliminating the need to write equations in a language or program. Written equations, however, should be used to describe the model in the Model Development step. The Model Implementation step can then proceed by constructing the model in Simulink. A new model window opens when clicking the **New model** icon or by choosing **File > New > model**. Block diagram models are built by connecting sinks, sources, math operators, and connectors from a comprehensive block library. To select simulation parameters, choose **Simulation > Configuration Parameters** in the **Model** toolbar menu. For example, you can set the **Start time** and **Stop time**. The variable-step integration methods described previously are also available in Simulink, as are several fixed-step solvers. To select a solver, in the **Configuration Parameters** window, select **Fixed-step** or **Continuous-step** in the **Type** field. Then select the **Solver**. Using scopes and other display blocks, you can see the simulation results while the simulation runs. You then can change many parameters and see what happens for "what if" exploration (Section 1.2.4). The simulation results can be put in the MATLAB workspace for postprocessing and visualization.

To start building a block diagram, begin by opening a **New model** window. Blocks can then be dragged from the library to the model window. Often it is better to begin the block diagram construction with the output, and then work backward to the input. Source blocks provide inputs to the model, such as a constant, a pulse generator, and a random number generator. Sink blocks in the Sink Library are blocks to display or save the results of the simulation. Two commonly used blocks are the **Scope** block, which displays the results in real time, and the **To Workspace** block, which saves the simulation output to the workspace. Math operators include multipliers, trigonometric functions, gain, sums, and summing point blocks. In the **Continuous** block library are the **Integrator** block, which integrates the signal entering the block, and the **Derivative** block, which takes the derivative of the entering signal.

An example of a block diagram model is the inhalation dose model described in Example 5.1. The **Model** window for this example is displayed in Figure 3.4. The differential equation describing this model is:

$$\frac{dU_L}{dt} = 10^{-6} Y \cdot V_T \cdot f - cU$$

where

Y = exposure concentration (mg·m^{-3}),
V_T = tidal volume (ml·breath^{-1}), and
f = breathing frequency (breaths·minute^{-1})
10^{-6} = constant to convert m^3 to ml
c   = elimination rate constant

We start the model with the scope sink block, which receives the output signal from the integrator block. The integrator block output is the state variable U. The output from the integrator block also feeds back to a summing block after being multiplied by the elimination rate constant. The negative sign in the summing block finishes construction of the negative term in Equation (3.1). The product of the parameters, Y, V_T, f, and the conversion constant, comprise the positive signal entering the summing block. This signal represents the positive term in Equation (3.1). These two terms now represent the whole differential equation, which is input to the integrator to solve the equation.

There are two ways of starting a simulation: (1) the **Simulation > start** menu in the **inhale Model** window, and (2) by clicking the **Start Simulation** icon on the **inhale Model** toolbar (Figure 3.5).

EXERCISES

1. Given matrix

$$A = \begin{bmatrix} 1 & 2 & 3 \\ 4 & 5 & 6 \end{bmatrix}$$

and matrix

$$B = \begin{bmatrix} 7 & 8 & 9 \\ 10 & 11 & 12 \end{bmatrix}$$

write a column vector C using columns 1 and 3 of A, column 2 of B, and column 3, row 1 of B.

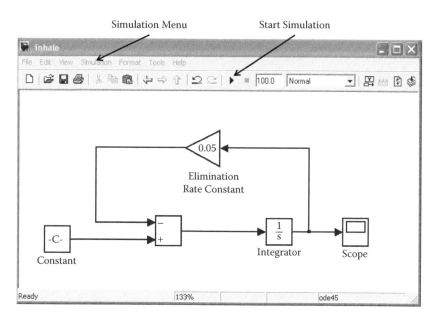

FIGURE 3.5   Block diagram for inhalation dose model.

2. Copy the text file named mercurydata.txt and save it to the MATLAB folder on the C drive (so the file can now be located at C:\MATLAB\mercurydata.txt).
3. Using the statements to read the text file, mercurydata.txt in Section 3.12, load the data from the text data files.
4. Run the m-files **resp4** with the function **resp3**. Check to see if your output matches Figure 5.9.
5. Copy the Excel file named mercurydata.xls and save it to the MATLAB folder just created (so the file can now be located at C:\MATLAB\mercurydata.xls).
6. Using the statement **[ndata,text]=xlsread('c:\MATLAB\mercurydata. xls','elimination'),** load the data from the Excel file (mercurydata.xls).
7. Define the four variables: Temperature, Day, Replicate, Hg concentration.
8. Now, save the file ndata keeping only the variables Day and Conc: save ('ndata','Day','Conc').
9. Run the m-files **resp4** with the function **resp3**. Check to see if your output matches Figure 5.9.
10. Run the Simulink model in Figure 3.5. Open Simulink and then the model file **inhale** to begin. Check to see if your output matches Figure 5.11.

REFERENCES

The MathWorks. 2010a. MATLAB® 7 *Getting Started Guide*. Natick, MA: The MathWorks, Inc.
The MathWorks. 2010b. Simulink® 7 *Getting Started Guide*. Natick, MA: The MathWorks, Inc.

4 Introduction to Stochastic Modeling

Deterministic modeling involves the use of parameters that take on only a single value, thus "determining" the model's outcome. Stochastic modeling, on the other hand, involves the use of random variables. Not only can model coefficients be functions of other variables, they can be functions of random variables and thus can be random variables themselves.

Definition 4.1: A *random variable* is a variable that takes on values with some relative frequency. ∎

In this way of classifying models, those with random (stochastic) variables are called *stochastic models* and those without are called *deterministic models*. Random variables are used to represent the random variation or *unexplained* variation in the state variables. Stochastic-differential equations can include random variables expressed either as random inputs (Equation [4.1]) or as parameters with a random error term (Equation [4.2]).

$$\frac{d^n y}{dt^n} + a_1 \frac{d^{n-1} y}{dt^{n-1}} + \cdots + a_{n-1} \frac{dy}{dt} + a_n y = f(t) + \varepsilon(t) \tag{4.1}$$

where $\varepsilon(t)$ is a random variable with mean 0 and variance σ^2, and

$$a_i = f_i(t) + \alpha_i(t) \tag{4.2}$$

where $\alpha_i(t)$ are normal random variables with mean 0 and variance σ^2.

4.1 INTRODUCTION TO PROBABILITY DISTRIBUTIONS

In a stochastic model, random variables representing input variables, model parameters (or both) will take on values according to some statistical distribution. In other words, there will be a probability associated with the value of the parameter or input variable. There is always uncertainty involved with the outcome of a random process. Over a certain time period an individual may or may not give birth, be exposed to a toxicant, or die. The probability that such events occur is a numerical value between 0 and 1, called the *probability mass*. We can define a function that assigns a probability to the random event as the *probability mass function* (PMF). A PMF often can be represented by a bar graph, or histogram, drawn over the sample space for the random variable (Figure 4.1).

Every event, x_0, defined by the random variable must have some probability, p_x, that must lie between 0 and 1 and the sum of all $p_x(x_0)$ must equal 1.0:

$$0 \le p_x(x_0) \le 1$$

$$\sum_{x_0} p_x(x_0) = 1 \tag{4.3}$$

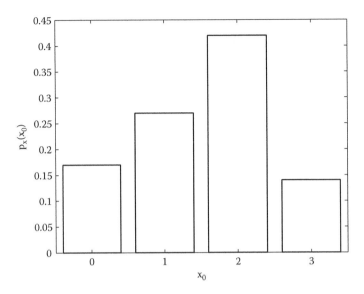

FIGURE 4.1   Probability mass function for random variable x.

The cumulative distribution function (CDF) for a random variable X is the probability that the random variable X takes on a value less than or equal to x:

$$P(X \leq x) = F(x) \tag{4.4}$$

It can be obtained from the PMF by noting that

$$F_x(x_0) = P(X \leq x_0) = \sum_{u \leq x_0} f(u) \tag{4.5}$$

where the sum on the right is taken over all values of u for which $u \leq x_0$ (Figure 4.2). The CDF can be sampled to obtain parameter values from a given distribution.

In most cases, state variables and parameters will be continuous variables, that is, those that can take on an infinite number of values. The probability that a continuous random variable takes on any particular value is zero. Therefore, for continuous random variable X, we define the probability that X lies between two different values a and b by the integral taken between a and b of the function $f(x)$:

$$P(a < X \leq b) = \int_a^b f(x)dx \tag{4.6}$$

where $f(x)$ is called the *probability density function* (PDF, Figure 4.3).

The properties of the continuous CDF are similar to the discrete CDF: the PDF has to be greater than or equal to zero and the integral of the PDF between minus infinity and plus infinity is 1. This is equivalent to the discrete probability mass function.

$$f(x) \geq 0$$

$$\int_{-\infty}^{\infty} f(x)dx = 1 \tag{4.7}$$

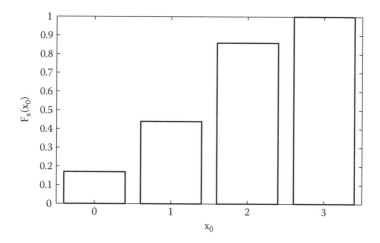

FIGURE 4.2 Example of a discrete cumulative distribution function for random variable x.

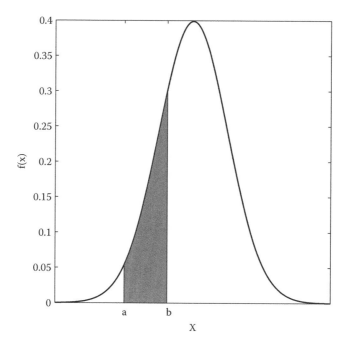

FIGURE 4.3 Example of a continuous probability density function.

The cumulative distribution function $F(x)$ (Figure 4.4) defines the range of possible values of the random variable x. Therefore, for the continuous random variable X, we define the probability that X lies between two different values a and b by the integral taken between a and b of the probability density function $f(x)$:

$$F(x) = P(a < X \leq b) = \int_{a}^{b} f(x)dx \qquad (4.8)$$

We obtain the probability by subtracting the probability of a from the probability of b in the CDF.

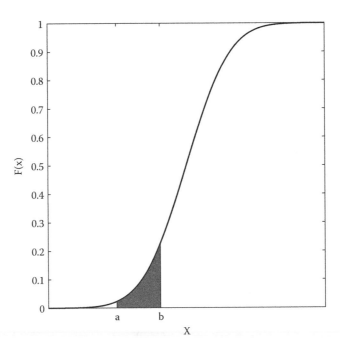

FIGURE 4.4   Example of a continuous cumulative distribution function.

In the distributions described above, the independent variable is plotted on the x axis and the probability on the y axis. In sampling a continuous distribution, however, we first obtain a probability on the x axis and then a value of the variable on the y axis, using what is called an *inverse distribution*. In the next section we examine several continuous and discrete probability distributions. In Monte Carlo simulation, we can sample these distributions to obtain random variates for parameter values (Section 4.4).

4.2   EXAMPLE PROBABILITY DISTRIBUTIONS

4.2.1   CONTINUOUS DISTRIBUTIONS

4.2.1.1   Uniform

Probability density function

$$f(x) = \begin{cases} \dfrac{1}{b-a} & a < x < b \\ 0 & \text{otherwise} \end{cases} \tag{4.9}$$

Cumulative distribution function

$$F(x) = \begin{cases} 0 & x < a \\ \dfrac{x-a}{b-a} & a \leq x \leq b \\ 1 & b < x \end{cases} \tag{4.10}$$

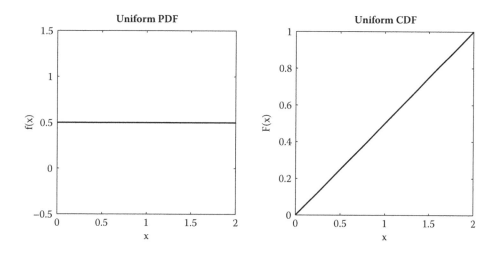

FIGURE 4.5 Probability density function and cumulative distribution function for uniform distribution.

The PDF is defined over the interval from a to b on the x axis. The probability along the interval is 1 over the length of the interval. The graphs for the PDF and CDF are shown in Figure 4.5. In Figure 4.5 the interval is $2 - 0 = 2$. Therefore, the probability is 1 over 2 or 0.5. The uniform distribution is a special case of the beta distribution when a and b equal 1.

4.2.1.2 Exponential

Probability density function

$$f(x) = \begin{cases} \dfrac{1}{\beta} e^{-\frac{x}{\beta}} & \text{if } x \geq 0 \\ 0 & \text{otherwise} \end{cases} \tag{4.11}$$

Cumulative distribution function

$$F(x) = \begin{pmatrix} 1 - e^{-\frac{x}{\beta}} & \text{if } x \geq 0 \\ 0 & \text{otherwise} \end{pmatrix} \tag{4.12}$$

The exponential distribution often is used to model one-compartment elimination. The graphs for the PDF and CDF are shown in Figure 4.6.

The parameter β is the mean of the distribution. The exponential PDF has no shape parameter, as it has only one shape. The exponential PDF is always convex, and is stretched to the right as it decreases in value.

4.2.1.3 Gamma

Probability density function

$$f(x) = \begin{cases} \dfrac{x^{a-1} e^{-\frac{x}{b}}}{\Gamma(a) b^a} & x > 0 \\ 0 & \text{otherwise} \end{cases} \tag{4.13}$$

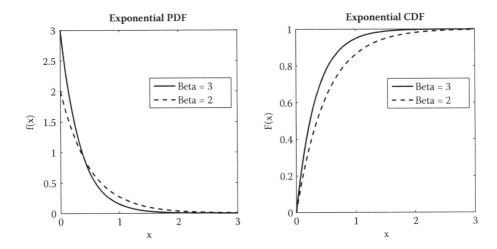

FIGURE 4.6 Probability density function and cumulative distribution function for exponential distribution. Dashed lines are beta = 2 and solid lines are beta = 3.

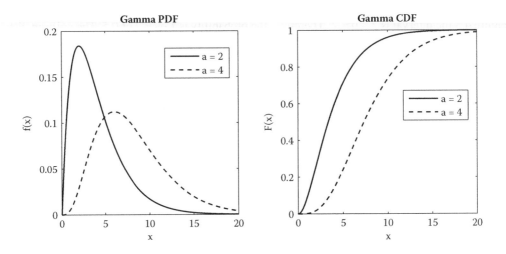

FIGURE 4.7 Probability density function and cumulative distribution function for the gamma distribution. The value of b is 2.0. Solid lines are $a = 2$, dashed lines are for $a = 4$.

Cumulative distribution function

$$F(x) = \begin{cases} \dfrac{1}{b^a \Gamma(a)} \displaystyle\int_0^x t^{a-1} e^{\frac{t}{b}} dt & x > 0 \\[2ex] 0 & \text{otherwise} \end{cases} \qquad (4.14)$$

The graphs for the PDF and CDF are shown in Figure 4.7. The exponential and chi-square functions are special cases of the gamma function. The parameter a is a shape parameter and must be greater than zero. The parameter b is a scale parameter, which also must be greater than zero. There is no known analytical solution to the integral Equation (4.14). To obtain random variates, the MATLAB function **gaminv** uses an iterative approach (Newton's method) to converge on the solution.

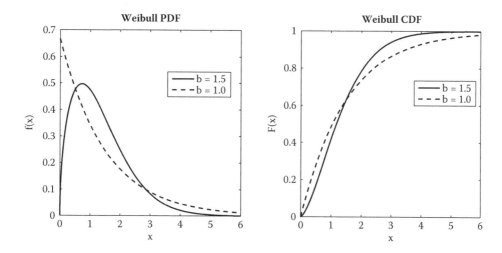

FIGURE 4.8 Probability density function and cumulative distribution function for the Weibull distribution. The value of a is 1.5. The solid curves are $b = 1.5$ and dashed curves are $b = 1.0$.

4.2.1.4 Weibull
Probability density function

$$f(x) = \begin{cases} ba^{-b}x^{b-1}e^{-(x/a)^b} & \text{if } x > 0 \\ 0 & \text{otherwise} \end{cases} \tag{4.15}$$

Cumulative distribution function

$$F(x) = \begin{cases} 1 - e^{-(x/a)^b} & \text{if } x > 0 \\ 0 & \text{otherwise} \end{cases} \tag{4.16}$$

The graphs of the Weibull PDF and CDF are shown in Figure 4.8. The exponential distribution is a special case of the Weibull distribution when b is equal to 1.

4.2.1.5 Normal
Probability density function

$$f(x) = \frac{1}{\sqrt{2\pi\sigma^2}} e^{\frac{-(x-\mu)^2}{2\sigma^2}} \qquad -\infty < x < \infty \tag{4.17}$$

Cumulative distribution function

$$F(x) = \frac{1}{\sigma\sqrt{2\pi}} \int_{-\infty}^{x} e^{\frac{-(t-\mu)^2}{2\sigma^2}} dt \tag{4.18}$$

Graphs of the normal PDF and CDF are shown in Figure 4.9. The mean μ is the location parameter and the standard deviation σ is the scale parameter. The *standard normal* distribution has $\mu = 0$ and $\sigma = 1$. If x is normal, then $y = (x-\mu)/\sigma$ is standard normal. As with the gamma distribution, the cumulative distribution function for a normal random variable cannot be found in closed form. The

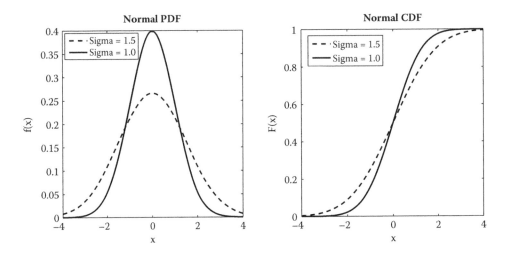

FIGURE 4.9 Probability density function and cumulative distribution function for the normal distribution. In these figures, the mean is zero. The solid lines have $\sigma = 1.0$ and the dashed lines have $\sigma = 1.5$.

normal inverse function, therefore, does not have a closed form. The MATLAB function **norminv** is calculated from the inverse cumulative error function **erfcinv**.

4.2.1.6 Lognormal
Probability density function

$$f(x) = \begin{cases} \dfrac{1}{x\sqrt{2\pi\sigma^2}}\, e^{\frac{-(\ln x - \mu)^2}{2\sigma^2}} & \text{if } x > 0 \\ 0 & \text{otherwise} \end{cases} \tag{4.19}$$

Cumulative distribution function

$$F(x) = \frac{1}{\sigma\sqrt{2\pi}} \int_0^x \frac{e^{\frac{-(\ln(t)-\mu)^2}{2\sigma^2}}}{t}\,dt \tag{4.20}$$

Graphs of the lognormal PDF and CDF are shown in Figure 4.10. There is no closed form of the CDF. The inverse lognormal function, therefore, has no closed form and can be estimated using the inverse cumulative error function as with the normal distribution. The MATLAB function **logninv** returns a value from the inverse lognormal distribution using **erfcinv**. The normal and lognormal distributions are closely related. If X is distributed lognormally with parameters μ and σ, then $\log(X)$ is distributed normally with mean μ and standard deviation σ.

4.2.1.7 Beta
Probability density function

$$f(x) = \begin{cases} \dfrac{x^{a-1}(1-x)^{b-1}}{B(a,b)} & \text{if } 0 < x < 1 \\ 0 & \text{otherwise} \end{cases} \tag{4.21}$$

where B is the Beta function.

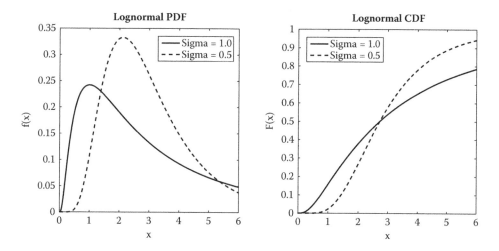

FIGURE 4.10 Probability density function and cumulative distribution function of the lognormal distribution. The value of μ is set equal to 1.0. The solid lines are $\sigma = 1.0$ and dashed lines are $\sigma = 0.5$.

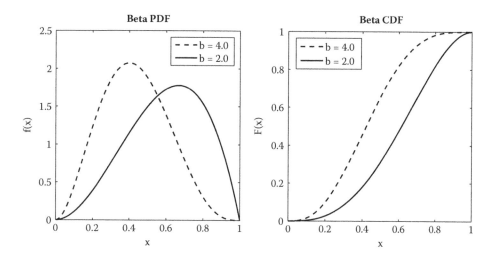

FIGURE 4.11 Probability density function and cumulative distribution function for the beta distribution. The value of a is set to 3.0. Solid lines are $b = 2.0$ and dashed lines are $b = 4.0$.

Cumulative distribution function

$$F(x) = \frac{1}{B(a,b)} \int_0^x t^{a-1} (1-t)^{b-1} dt \tag{4.22}$$

Graphs of the beta PDF and CDF are shown in Figure 4.11. The parameters a and b must be positive, and the values in x must lie on the interval [0, 1]. There is no closed form to the CDF. To obtain random variates, the MATLAB **betainv** function uses Newton's method with modifications to constrain steps to the allowable range for x, i.e., [0 1].

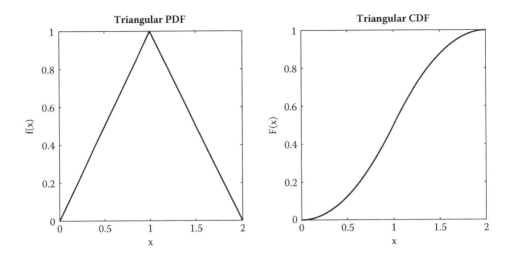

FIGURE 4.12 Probability density function and cumulative distribution function for the triangular distribution. The parameter $b = 2$, $a = 0$, and $c = 1$.

4.2.1.8 Triangular
Probability density function

$$f(x) = \begin{cases} \dfrac{2(x-a)}{(b-a)(c-a)} & \text{if } a \le x \le c \\ \dfrac{2(b-x)}{(b-a)(b-c)} & \text{if } c < x \le b \\ 0 & \text{otherwise} \end{cases} \tag{4.23}$$

Cumulative distribution function

$$F(x) = \begin{cases} 0 & \text{if } x \le a \\ \dfrac{(x-a)^2}{(b-a)(c-a)} & \text{if } a \le x \le c \\ 1 - \dfrac{(b-x)^2}{(b-a)(b-c)} & \text{if } c < x \le b \\ 1 & \text{if } b < x \end{cases} \tag{4.24}$$

Graphs of the triangular PDF and CDF are shown in Figure 4.12.

4.2.1.9 Logistic
The logistic distribution was developed by the doctor and mathematician, Pierre François Verhulst, while working on demography in the early 1800s (Verhulst 1838, 1845).
    Probability density function

$$f(x) = \frac{2.2\, y(1-y)}{P_3} \tag{4.25}$$

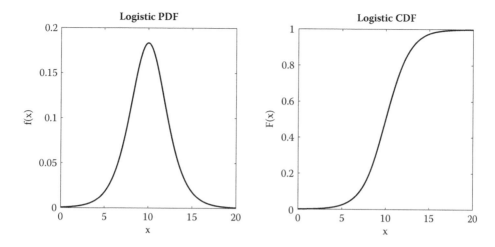

FIGURE 4.13 Probability density function and cumulative distribution function for the logistic distribution.

Cumulative distribution function

$$F(x) = \frac{1}{1 + e^{(2.2/P_3)(P_2 - x)}}$$

(4.26)

Graphs of the logistic PDF and CDF are shown in Figure 4.13.

Example 4.1

We know that the distribution of deaths resulting from exposure to a toxicant can have a normal probability distribution, although the exposure concentrations may have to be log transformed (see Section 5.1.1.1). This dose response often is characterized by a single value: the dose at which 50% of the exposed individuals die, the LD_{50}. The logistic distribution can be used in place of the normal distribution in most cases. In Chapter 7, we give an example of using nonlinear least squares regression to estimate the parameters in the logistic model. In this example, we used the same logistic model to estimate the LD_{50} of tadpoles exposed to zinc concentrations (Gottschalk 1995). The value of P_1 is 1 if there is 100% mortality. The two fitted parameters are P_2, the LD_{50}, and P_3, the range of dose in which mortality increases from 10% to 50%. A plot of the data and the fitted curve are shown in Figure 4.14.

The estimated mean and standard deviation values from the regression are 10.00 ± 0.4664 for P_2 and 2.34 ± 1.0117 for P_3. It is often better to fit the CDF than the PDF because there are fewer inflection points in the CDF and it is easier to obtain cumulative mortality data than actual mortality rates. The PDF is obtained by taking the derivative of PDF. The graph of the PDF is shown in Figure 4.15. The value of P_1 is 1 if there is 100% mortality.

$$\frac{dF(x)}{dx} = \frac{2.2F(P_1 - F)}{P_1 \cdot P_3}$$

(4.27)

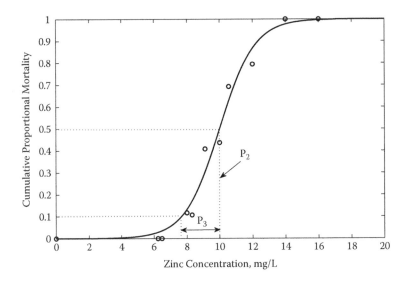

FIGURE 4.14   Cumulative proportional mortality in tadpoles as a function of zinc concentration. (Data from Jennifer Gottschalk, "Copper and Zinc Toxicity to the Gray Treefrog (*Hyla chrysocelis*) and the Northern Leopard Frog (*Rana pipiens*)," Master of Science thesis, Clemson University, 1995.)

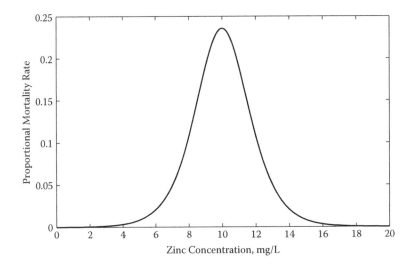

FIGURE 4.15   Plot of the PDF of tadpole mortality.

4.2.2   DISCRETE DISTRIBUTIONS

4.2.2.1   Bernoulli

$$p(x) = \begin{cases} 1-p & \text{if } x = 0 \\ p & \text{if } x = 1 \\ 0 & \text{otherwise} \end{cases} \tag{4.28}$$

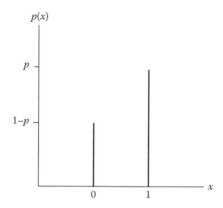

FIGURE 4.16   Bernoulli probability mass function.

The graph of the Bernoulli probability mass function is shown in Figure 4.16. The Bernoulli distribution has two possible outcomes, one, with a probability of p and zero, with a probability of $1-p$. We sometimes refer to the random variable x as a Bernoulli random variable. A single sample from the Bernoulli PMF is called a *Bernoulli trial*. We can consider an outcome of a Bernoulli trial as a success if the value is 1 and a failure if the value is 0. A series of independent Bernoulli trials, each with the same probability of success, is a *Bernoulli process*.

4.2.2.2   Binomial

Probability mass function

$$p(x) = \begin{cases} \binom{n}{x} p^x (1-p)^{n-x} & x = 0,1,2,\ldots,n \\ 0 & \text{otherwise} \end{cases} \tag{4.29}$$

where $\binom{n}{x}$ is the binomial coefficient defined by

$$\binom{n}{x} = \frac{n!}{x!(n-x)!}$$

Cumulative distribution function

$$F(x) = \begin{cases} 0 & x < 0 \\ \sum_{i=0}^{\lfloor x \rfloor} \binom{n}{i} p^i (1-p)^{n-i} & \text{if } 0 \le x \le n \\ 1 & \text{if } n < x \end{cases} \tag{4.30}$$

Graphs of the binomial PMF and CDF are shown in Figure 4.17. The MATLAB binomial inverse CDF function, $\mathbf{x = binoinv(y,n,p)}$, returns the smallest integer $\mathbf{x}$ such that the binomial CDF evaluated at $\mathbf{x}$ is equal to or exceeds $\mathbf{y}$. $\mathbf{y}$ is the probability of observing $\mathbf{x}$ successes in $\mathbf{n}$ independent trials where $\mathbf{p}$ is the probability of success in each trial. Each $\mathbf{x}$ is a positive integer less than or equal to $\mathbf{n}$.

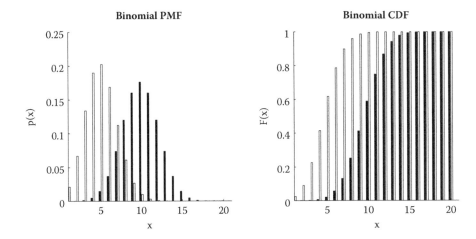

FIGURE 4.17   Probability mass function and cumulative distribution function for the binomial distribution. Clear bars are $p = 0.25$ and solid bars are $p = 0.50$.

4.2.2.3   Discrete Uniform

The discrete uniform distribution is a simple distribution that puts equal weight on the integers between i and j. Usually, i takes on a value of 1, so the range is from 1 to j.

Probability mass function

$$p(x) = \begin{cases} \dfrac{1}{j-i+1} & x = i, i+1, \ldots, j \\ 0 & \text{otherwise} \end{cases} \tag{4.31}$$

Cumulative distribution function

$$F(x) = \begin{cases} 0 & \text{if } x < i \\ \dfrac{\lfloor x \rfloor - i + 1}{j - i + 1} & \text{if } i \le x \le j \\ 1 & \text{if } j < x \end{cases} \tag{4.32}$$

where $\lfloor x \rfloor$ is the largest integer $\le x$.

The result, $F(x)$, is the probability that a single observation from the discrete uniform distribution with maximum j will be a positive integer less than or equal to x. Graphs of the PMF and CDF for the discrete uniform distribution are shown in Figure 4.18. In this figure, i is set to 1 and j equals 10, so $p(x)$ will equal 0.1 for all values of x from 1 to 10.

4.2.2.4   Geometric

Probability mass function

$$p(x) = \begin{cases} p(1-p)^x & x = 0, 1, 2, \ldots \\ 0 & \text{otherwise} \end{cases} \tag{4.33}$$

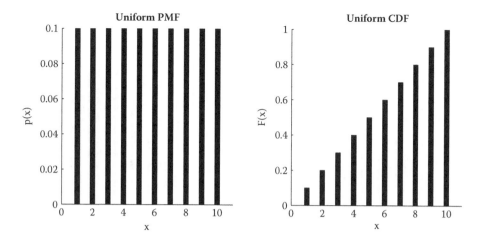

FIGURE 4.18 Probability mass function and cumulative distribution function for the discrete uniform distribution.

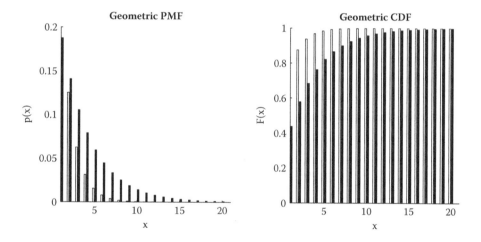

FIGURE 4.19 Probability mass function and cumulative distribution function for the geometric distribution. Clear bars are $p = 0.5$ and solid bars are $p = 0.25$.

The parameter p must lie on the interval [0 1].

Cumulative distribution function

$$F(x) = \sum_{i=0}^{\lfloor x \rfloor} p(1-p)^i \tag{4.34}$$

The geometric distribution results from a series of Bernoulli trials. The random variable x can be defined as the number of trials it takes to reach the first success and has a geometric PMF.

Graphs of the geometric PMF and CDF are shown in Figure 4.19.

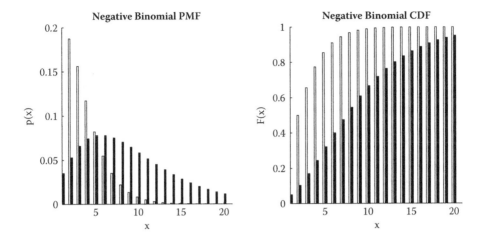

FIGURE 4.20 Probability mass function and cumulative distribution function for the negative binomial distribution. Clear bars are $p = 0.5$ and solid bars are $p = 0.25$.

4.2.2.5 Negative Binomial

Probability mass function

$$p(x) = \begin{cases} \begin{pmatrix} r + x - 1 \\ x \end{pmatrix} p^r (1 - p)^x & x = 0, 1, 2, \ldots \\ 0 & \text{otherwise} \end{cases} \tag{4.35}$$

Cumulative distribution function

$$F(x) = \begin{cases} \displaystyle\sum_{i=0}^{\lfloor x \rfloor} \begin{pmatrix} r + i - 1 \\ i \end{pmatrix} p^r (1 - p)^i & \text{if } x \geq 0 \\ 0 & \text{otherwise} \end{cases} \tag{4.36}$$

Graphs of the negative binomial PMF and CDF are shown in Figure 4.20.

4.2.2.6 Poisson

Probability mass function

$$p(x) = \begin{cases} \dfrac{e^{-\lambda} \lambda^x}{x!} & x = 0, 1, \ldots \\ 0 & \text{otherwise} \end{cases} \tag{4.37}$$

Cumulative distribution function

$$F(x) = \begin{cases} 0 & \text{if } x < 0 \\ e^{-\lambda} \displaystyle\sum_{i=0}^{\lfloor x \rfloor} \dfrac{\lambda^i}{i!} & \text{if } 0 \leq x \end{cases} \tag{4.38}$$

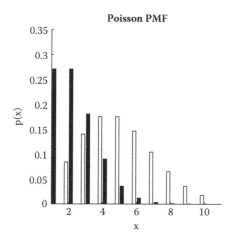
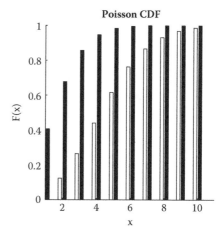

FIGURE 4.21 Probability mass function and cumulative distribution function for the Poisson distribution. Clear bars are for $\lambda = 5$; solid bars are for $\lambda = 2$.

The Poisson distribution has a single parameter, the mean and variance, λ. Graphs of the Poisson PMF and CDF are shown in Figure 4.21.

Example 4.2

The Bernoulli probability distribution was used to model the ingestion of pesticide granules by birds feeding in corn fields (Solomon et al. 2001). The number of granules ingested in an hour was considered a Bernoulli trial in which a granule was ingested with probability p:

$$p(x) = \begin{cases} 1-p & \text{if } x = 0 \\ p & \text{if } x = 1 \\ 0 & \text{otherwise} \end{cases} \tag{4.39}$$

In other words, if a bird consumed a granule, the Bernoulli random variable equals one, and if it didn't consume a granule, the variable equals zero. We used data from Fischer and Best (1995) that describe granule consumption on a daily basis, to test the Bernoulli model. Using the fact that a sum of a series of Bernoulli trials will comprise a binomial distribution, we compared a daily summation of granules ingested with the data from Fischer and Best.

The MATLAB m-file we used to generate the binomial distribution from a series of Bernoulli trials is:

```
% program bernoulli - generates sum of Bernoulli random variates
% and plots a histogram of the sums

p=.05;      % probability of success in a single trial
n=10;       % number of trials in a sum
m=1000;     % number of sums calculated
s = RandStream.create('mt19937ar','seed',sum(100*clock));
         RandStream.setDefaultStream(s);
% generate series of Bernoulli trials
for j = 1:m
   for i = 1:n
```

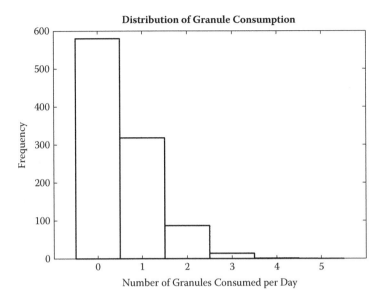

FIGURE 4.22   Binomial distribution formed by summing a series of Bernoulli trials.

```
        U(i)=rand;          % get uniform random variates
        if   U(i)<p
            X(i)=1;
        else X(i)=0;
        end
    end
    Y(j)=sum(X);            % calculate sum of Bernoulli trials
                           % to get binomial random variable
end

bins=[0 1 2 3 4 5]
hist(Y,bins)               % plot histogram of binomial distribution
xlabel('X')
ylabel('Frequency')
title('Binomial Distribution')
```

The graphic output from this m-file is shown in Figure 4.22.

4.2.3 EMPIRICAL DISTRIBUTIONS

Not every random variable can be modeled with a known distribution. There may not be enough evidence to determine the underlying distribution or there may not be an appropriate distribution. In this case, empirical distributions are often used. We base the empirical distribution on the outcomes of an experiment or experiments. The random variable is divided into classes and the number of occurrences of the random variable in each class is observed. The ratio of the number of outcomes, n_i, in each class i to the total number of observations in that class are defined as *relative frequencies*, f_i:

$$f_i = \frac{n_i}{\sum_i n_i} \tag{4.40}$$

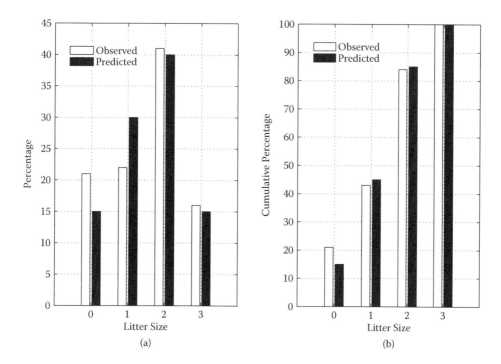

FIGURE 4.23   Empirical distribution of litter size: (a) probability mass function, (b) cumulative distribution function. (Adapted from K. R. Dixon, T.-Y. Huang, K. T. Rummel, et al. An Individual-Based Model for Predicting Population Effects from Exposure to Agrochemicals. In *Challenges in Applied Population Biology, Aspects of Applied Biology* 53, ed. M. B. Thomas and T. Kedwards, 241–251. Wellesbourne, Warwick, UK: Association of Applied Biologists, % Horticulture Research International, 1999.)

Example 4.3

An example of a discrete empirical distribution is an empirical distribution used to describe the reproduction in an individual-based population model (Dixon et al. 1999). The probability of producing from zero to three offspring is defined (Figure 4.23a). The actual sampling is based on the cumulative distribution function (Figure 4.23b). If the sample value of a random variate from a uniform probability distribution is between 0 and 0.15, the number of offspring would be zero; if the value was between 0.15 and 0.45, the number of offspring would be one; and so on. The observed values that resulted from the sampling are within 10% of the input probabilities (Figure 4.23). The value of n represents the number of females in the given age class summed over time. In other words, it represents the number of decisions made concerning the number of offspring per female over the period of the simulation.

4.3   DISCRETE-STATE MARKOV PROCESSES

As we described in Chapter 2, systems can be defined as either *discrete* or *continuous* in terms of the time step used in defining the process. We also can define a system by whether the states of the system are discrete or continuous. A *discrete state process* is one in which the system can be in one of a set of mutually exclusive states that can change at discrete points in time according to a set of probabilistic rules called *state transition probabilities*. If a series of state transition probabilities are such that the probability of the system being in state j given only that it was in state i during the previous time step, the series satisfies the Markov condition and the process is termed a *discrete-state discrete-transition Markov process*.

In mathematical terms, the transition probabilities, p_{ij}, can be defined as:

$$p_{ij} = P[S_j(n) \mid S_i(n-1)] \tag{4.41}$$

where $S_j(n)$ is state j at time step n and $S_i(n-1)$ is state i at time step $(n-1)$. The probabilities must satisfy the following conditions: they are between 0 and 1, $0 \le p_{ij} \le 1$, and they must sum to 1 over states j, that is, for each row, the sum of the elements in that row equals 1

$$\sum_j p_{ij} = 1$$

The transition probabilities often are expressed as an $m \times m$ matrix:

$$[p] = \begin{pmatrix} p_{11} & p_{12}\cdots & p_{1m} \\ p_{21} & p_{22}\cdots & p_{2m} \\ \vdots & \vdots & \vdots \\ p_{m1} & p_{m2}\cdots & p_{mm} \end{pmatrix} \tag{4.42}$$

The initial numbers, or proportions, in each state are contained in an initial state vector:

$$[S_i(0)] = \begin{bmatrix} S_1(0) & S_2(0) & \cdots & S_m(0) \end{bmatrix} \tag{4.43}$$

To obtain the numbers in each state at some time n, we postmultiply the initial state vector by the transition matrix raised to the nth power:

$$[S_i(n)] = [S_i(0)] \times [p]^n = \begin{bmatrix} S_1(0) & S_2(0) & \cdots & S_m(0) \end{bmatrix} \times \begin{pmatrix} p_{11} & p_{12}\cdots & p_{1m} \\ p_{21} & p_{22}\cdots & p_{2m} \\ \vdots & \vdots & \vdots \\ p_{m1} & p_{m2}\cdots & p_{mm} \end{pmatrix}^n \tag{4.44}$$

making sure that the number of states in the initial state vector equals the number of transition probabilities in each column of the p matrix.

An alternative model structure to the matrix multiplication is a set of state equations obtained by expanding the matrices, that is, multiplying each state by its corresponding column of transition probabilities:

$$S_1(n+1) = S_1(n)p_{11} + S_2(n)p_{21} + \cdots + S_m(n)p_{m1}$$

$$S_2(n+1) = S_1(n)p_{12} + S_2(n)p_{22} + \cdots + S_m(n)p_{m2}$$

$$\vdots \quad = \quad \vdots \qquad \vdots \qquad\qquad \vdots \tag{4.45}$$

$$S_m(n+1) = S_1(n)p_{1m} + s_2(n)p_{2m} + \cdots + S_m(n)p_{mm}$$

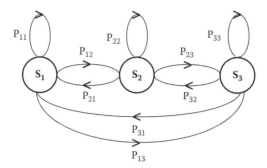

FIGURE 4.24   General form of state-transition diagram.

The Markov process can be pictured as a state-transition diagram showing the different states and their transition probabilities. For example, in a three-state Markov process, we would have the diagram shown in Figure 4.24:

Example 4.4

Field observation of deer movements shows that deer feed in three different habitats, S1, S2, and S3. Observation also shows the probabilities of a deer moving among the three habitats. We want to determine the transition probabilities and average time spent in each habitat after k time steps. The state-transition diagram (Figure 4.25) shows the states (habitats) and the transition probabilities.

These probabilities can be displayed as a transition matrix:

$$[p] = \begin{bmatrix} 0.6 & 0.2 & 0.2 \\ 0.3 & 0.4 & 0.3 \\ 0.1 & 0.2 & 0.7 \end{bmatrix} \tag{4.46}$$

Each row represents the probabilities of moving from a habitat to the habitat in each column. For example, a deer in habitat S1 has a 0.6 probability of staying in habitat S1, a 0.2 probability of moving to habitat S2, and a 0.2 probability of moving to habitat S3. Note also that each row sums to 1.0 as there is a 100% probability of a deer being in some habitat in the next time step. Let's assume that the deer starts in habitat S2, so the initial probability state vector is [0 1 0].

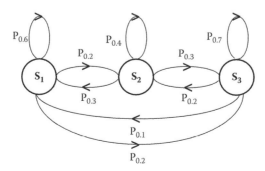

FIGURE 4.25   State-transition diagram for deer feeding habitats.

If the probabilities remain constant, we can obtain the k-step transition probabilities, p_{ij}, by using matrix multiplication to multiply $[p]$ by itself k times. For example, at the second time step where $k = 2$, the p matrix is squared:

$$[p_2] = \begin{bmatrix} 0.6 & 0.2 & 0.2 \\ 0.3 & 0.4 & 0.3 \\ 0.1 & 0.2 & 0.7 \end{bmatrix} \times \begin{bmatrix} 0.6 & 0.2 & 0.2 \\ 0.3 & 0.4 & 0.3 \\ 0.1 & 0.2 & 0.7 \end{bmatrix} = \begin{bmatrix} 0.6 & 0.2 & 0.2 \\ 0.3 & 0.4 & 0.3 \\ 0.1 & 0.2 & 0.7 \end{bmatrix}^2 = \begin{bmatrix} 0.4400 & 0.2400 & 0.3200 \\ 0.3300 & 0.2800 & 0.3900 \\ 0.1900 & 0.2400 & 0.5700 \end{bmatrix} \quad (4.47)$$

Now, to get the $k = 2$ probability state vector, we multiply the initial probability state vector by the $k = 2$ transition matrix:

$$P(2) = \begin{bmatrix} 0 & 1 & 0 \end{bmatrix} \times \begin{bmatrix} 0.4400 & 0.2400 & 0.3200 \\ 0.3300 & 0.2800 & 0.3900 \\ 0.1900 & 0.2400 & 0.5700 \end{bmatrix} = \begin{bmatrix} 0.3300 & 0.2800 & 0.3900 \end{bmatrix} \quad (4.48)$$

The results show the average time spent in each habitat after two time steps. The same procedure is used to find the transition probabilities and the probability state vector after many time steps. For example, for $k = 15$:

$$[p_{15}] = \begin{bmatrix} 0.6 & 0.2 & 0.2 \\ 0.3 & 0.4 & 0.3 \\ 0.1 & 0.2 & 0.7 \end{bmatrix}^{15} = \begin{bmatrix} 0.3000 & 0.2500 & 0.4500 \\ 0.3000 & 0.2500 & 0.4500 \\ 0.3000 & 0.2500 & 0.4500 \end{bmatrix} \quad (4.49)$$

$$P(15) = \begin{bmatrix} 0 & 1 & 0 \end{bmatrix} \times \begin{bmatrix} 0.3000 & 0.2500 & 0.4500 \\ 0.3000 & 0.2500 & 0.4500 \\ 0.3000 & 0.2500 & 0.4500 \end{bmatrix} = \begin{bmatrix} 0.3000 & 0.2500 & 0.4500 \end{bmatrix} \quad (4.50)$$

Two points need to be made. First, as k increases, the probabilities, p_{ij}, will converge to constant values. Second, the results are independent of the initial state vector. That is, the deer could have started out in habitat S1 or S3, and the final transition probabilities and probability state vector will be the same as those for S2. The MATLAB program used to generate the values above is as follows:

```
% Markov chain model of deer habitat use
p0=[0.6 0.2 0.2
    0.3 0.4 0.3
    0.1 0.2 0.7];        % transition probability matrix
S0=[0 1 0];              % vector of initial states
k=15                     % number of days
patk=p0^k                % transition probabilities
Satk=S0*patk             % average percentage of time spent
                         % in each habitat type
```

The output in the Command Window is for $k = 15$:

```
k =
    15
patk =
    0.3000    0.2500    0.4500
    0.3000    0.2500    0.4500
    0.3000    0.2500    0.4500
 Satk =   0.3000    0.2500    0.4500
```

4.4   MONTE CARLO SIMULATION

Monte Carlo is a numerical technique of finding a solution to a stochastic model. The process has been described by Dixon and Martin (2006). For those random features of the model, values are chosen from a probability distribution. Repeated runs of the model result in different outcomes. Parameter values may or may not come from a known probability distribution, such as uniform, normal, exponential, and so on. If the distribution is known, with known mean μ and variance σ^2, a parameter value P is determined by

$$P = \mu + R\sigma$$

A probability distribution can be calculated for a state variable in the model along with its mean and variance. Suppose the model has parameters $p_1, p_2, p_3, \ldots, p_n$, the state variable will be a function of the n parameters

$$x = f(p_1, p_2, p_3, \ldots, p_n) \tag{4.51}$$

Now we calculate a value for each parameter by sampling from its individual distribution function. We then obtain a value for the state variable x by running a simulation of the model. We repeat the process until we have n values of the state variable x. Finally, we determine the mean μ and variance σ^2 for x (Chapter 9).

To sample the parameter distribution, we actually use the inverse of the cumulative distribution function. In most cases we can use the inverse transform method. For some distributions, such as the normal, lognormal, gamma, and beta distributions, there is no closed form of the CDF and therefore iterative methods must be used. Fortunately, MATLAB has functions to sample most common continuous and discrete distributions, including those that require iterative methods.

To illustrate the inverse transform method, consider the exponential distribution (Figure 4.6). The probability density function for the exponential distribution is defined by:

$$f(x) = \begin{cases} \dfrac{1}{\beta} e^{\beta} & \text{if } x \geq 0 \\ 0 & \text{otherwise} \end{cases} \tag{4.52}$$

where β is the mean of the distribution. The CDF is

$$F(x) = \left(\begin{array}{ll} 1 - e^{-\frac{x}{\beta}} & \text{if } x \geq 0 \\ 0 & \text{otherwise} \end{array} \right) \tag{4.53}$$

In general, if Y is a function of X, that is, $Y = F(x)$, the inverse distribution is found by setting $X = F^{-1}(Y)$. To find F^{-1} for the exponential distribution, we set $u = F(x)$ and solve for x to obtain

$$F^{-1}(u) = -\beta \ln(1-u) \tag{4.54}$$

To generate the random variate x, we first generate a random variate u from a uniform distribution $U(0,1)$. The second step is to return $x = -\beta \ln(1 - U)$.

Example 4.5

This example follows Example 4.1 where the logistic equation was used to describe a dose response. The estimated means for parameters P_2 and P_3 were used in describing the model and the plots of the logistic PDF and CDF (Figures 4.14 and 4.15). We know, however, that these estimates have variability represented by their standard deviations. The results of least squares regression yield 95% half-confidence intervals. The standard deviations can be estimated by

$$ULC = \bar{x} + t_{.05}\, s_{\bar{x}}$$

$$= \bar{x} + t_{.05}\, \frac{s}{\sqrt{n}} \tag{4.55}$$

$$s = \frac{(UCL - \bar{x})\sqrt{n}}{t_{.05}}$$

where
  UCL = upper confidence limit
  $\bar{x}$ = sample mean of parameter
  $t_{.05}$ = student's t for two-tailed distribution with 95% confidence
  $s_{\bar{x}}$ = standard error
  s = standard deviation
  n = sample size (number of observations in regression)

In this example, the standard deviation of P_2 is estimated by

$$s = \frac{(10.3133 - 10.000)\sqrt{11}}{2.228} \tag{4.56}$$

$$= 0.4664$$

Likewise, the estimate of the standard deviation for P_3 is 1.0117. A plot of the distributions for P_2 and P_3 is shown in Figure 4.26.

In a model that includes a dose–response function such as Equation (5.4), we need to obtain estimates of parameters P_2 and P_3. We used the MATLAB function **norminv** in the m-file **histogramoverlay2** to obtain the parameter samples from the distributions in Figure 4.26. To verify that we are sampling from normal distributions for P_2 and P_3, we generated 10,000 random variates, plotted a histogram, and overlaid the PDFs (Figure 4.27). The results show a close agreement between the actual PDFs and the distributions of sampled parameter values.

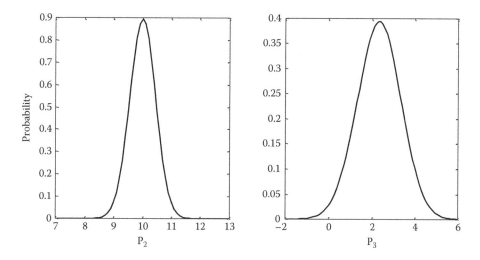

FIGURE 4.26 Plots of normal PDFs for parameters P_2 and P_3.

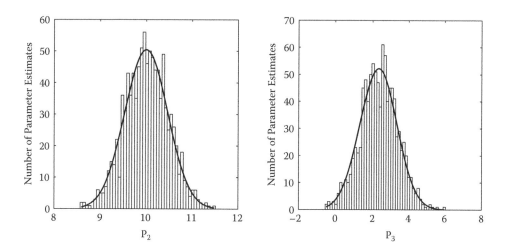

FIGURE 4.27 Plot of samples of parameters P_2 and P_3 from inverse CDFs overlaid with PDFs.

EXERCISES

1. Using m-file **doseresponse**, find the probability of a response at doses between 10 and 20.
2. Write a program to draw a sample from the Weibull inverse CDF and plot the histogram overlaid with the Weibull PDF. Use the parameters: $a = 1.5$ and $b = 1.5$. Hint: Use the m-file **histogramoverlay2** as a starting point.
3. In Example 4.2, we saw that the binomial distribution describes the ingestion of pesticide granules by birds picking up grit. Assume that the probability of a bird ingesting a granule on any trial of picking up grit is 0.05. If a bird picks up 100 items in a day, how many pesticide granules will it consume if the probability of observing that number (drawn from a uniform PDF) has a probability of at least 0.5?

REFERENCES

Dixon, K. R., and C. L. Martin. 2006. "Probabilistic Risk Assessment." In *Endocrine Disruptors: Biological Basis for Health Effects in Wildlife and Humans*, eds. J. A. Carr and D. O. Norris, 188–201. New York: Oxford Press.

Dixon, K. R., T.-Y. Huang, K. T. Rummel, et al. 1999. "An Individual-Based Model for Predicting Population Effects from Exposure to Agrochemicals." In *Challenges in Applied Population Biology, Aspects of Applied Biology* 53, ed. M. B. Thomas and T. Kedwards, 241–251. Wellesbourne, Warwick, UK: Association of Applied Biologists, % Horticulture Research International.

Fischer, D. L., and L. B. Best. 1995. "Avian Consumption of Blank Pesticide Granules Applied at Planting to Iowa Cornfields." *Environmental Toxicologu and Chemistry* 14:1543–1549.

Gottschalk, J. A. 1995. "Copper and Zinc Toxicity to the Gray Treefrog (*Hyla chrysocelis*) and the Northern Leopard Frog (*Rana pipiens*)." Master's thesis, Clemson University.

Solomon, K. R., J. P. Giesy, R. J. Kendall, et al. 2001. "Chlorpyrifos: Ecotoxicological Risk Assessment for Birds and Mammals in Corn Agroecosystems." *Human and Ecological Risk Assessment* 7:497–632.

Verhulst, P.-F. 1838. "Notice sur la loi que la population suit dans son accroissement." *Correspondance Mathématique et Physique* 10:113–121.

Verhulst, Pierre-François. 1845. "Recherches mathématiques sur la loi d'accroissement de la population." *Nouveaux Mémoires de l'Académie Royale des Sciences et Belles-Lettres de Bruxelles* 18:1–42.

5 Modeling Ecotoxicology of Individuals

5.1 TOXIC EFFECTS ON INDIVIDUALS

5.1.1 THE DOSE–RESPONSE RELATIONSHIP

The dose–response relationship defines the functional relation between the exposure concentration received by an organism and the resulting effect caused by the exposure. This relationship has played a major role in the history of toxicology, primarily in the calculation of the median lethal dose (LD_{50}), the dose that causes 50% mortality in a population. The dose–response relationship has a significant role in the application of modeling to ecotoxicology as well. We can define two types of response: (1) a quantal response, in which the response occurs or it doesn't (i.e., death), and (2) a graded response in which the magnitude of response increases or decreases as the dose level increases.

5.1.1.1 Quantal Response

Most work in the past has been on quantal response studies (Brown 1978, Finney 1964) and we will examine this type of response in a modeling context first. The frequency distribution of responses typically is skewed to the right (Figure 5.1) because there usually are a small number of individuals that are resistant to even high doses.

Other responses may have a more normal distribution (Figure 5.2). Historically, skewed distributions were transformed using the logarithm of dose to produce a normal distribution. This was necessary in the early calculations of the median lethal dose because of limited computing power. This transformation is not required for modeling, however, and in fact is no longer necessary for LD_{50} calculations.

We can consider dose to be a continuous random variable because values may fall anywhere within a continuous range from zero to infinity. The appropriate probability measure, therefore, is the probability density function $f(D)$ for random variable D, defined by

$$\text{Prob}(a < D \leq b) = \int_a^b f(D)\,dD \tag{5.1}$$

It should be pointed out that the probability of a dose is defined over a dose interval by the integral over that interval and not by the density function itself. This is because the probability of a single exact dose is zero when dose is defined as a continuous random variable. For example, in Figure 5.3, the shaded area represents the proportion of the population exhibiting a response between dose levels a and b.

One of the properties of the probability density function is that when integrated over all possible dose values, the probability equals 1:

$$\int_0^\infty f(D)\,dD = 1 \tag{5.2}$$

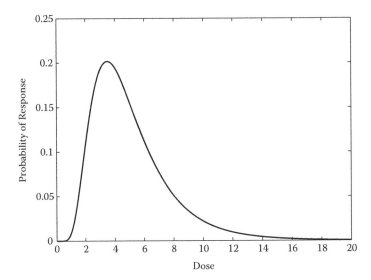

FIGURE 5.1   Dose–response showing skewed response.

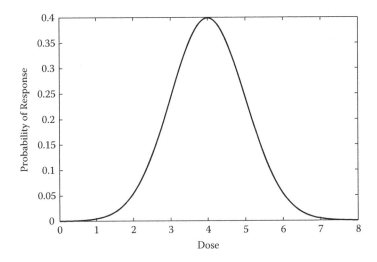

FIGURE 5.2   Dose–response where the probability of response is normally distributed as a function of dose.

This is equivalent to the assumption that a high enough dose level exists that all members of the population will respond.

We will find it more convenient to work with the integral of the probability density function, called the cumulative distribution function.

$$p(D_0) = \text{Prob}(D \le D_0) = \int_0^{D_0} f(D)\,dD \tag{5.3}$$

To find the proportion of the population that responds to doses between two levels, a and b, we now can subtract the proportion responding at the lower level a from the proportion responding

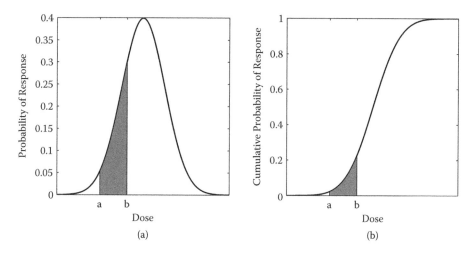

FIGURE 5.3 Dose–response showing probability of response at doses between *a* and *b*. Figure 5.3a is probability of response (PDF) and Figure 5.3b is the cumulative response (CDF).

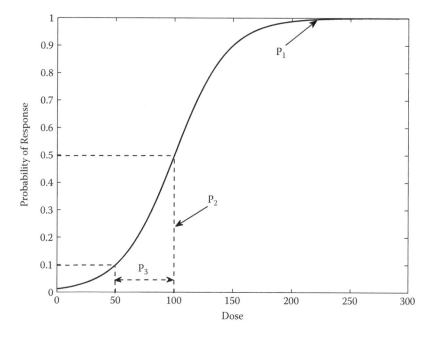

FIGURE 5.4 Dose–response generated from logistic function (Equation [5.4]).

at the higher level *b*, giving the probability that an individual will respond to a dose within this interval. By setting the lower level equal to zero, we can calculate the probability of an individual responding to a given dose (Figure 5.4).

In this figure, the response, $F(Q)$, is described by the logistic function:

$$F(Q) = \frac{P_1}{1 + e^{(2.2/P_3)(P_2 - Q)}}$$

(5.4)

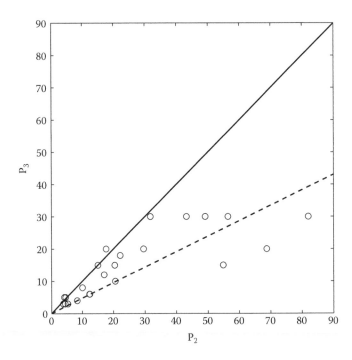

FIGURE 5.5 Plot of the relationship between parameters P_2 and P_3. The solid line is where P_2 equals P_3. The dashed line is the regression of P_3 on P_2 with a zero intercept. (Data from Jennifer Gottschalk, "Copper and Zinc Toxicity to the Gray Treefrog (*Hyla chrysocelis*) and the Northern Leopard Frog (*Rana pipiens*)," Master of Science thesis, Clemson University, 1995.)

where
P_1 = maximum response
P_2 = LD_{50}
P_3 = difference between LD_{10} and LD_{50}
Q = dose or exposure concentration

There is a relationship between the parameters P_2 and P_3 (Figure 5.5). The parameter P_3 has to be less than P_2 and should define the dose–response curve at the point (0,0). In other words, there has to be zero mortality probability at a zero dose. A regression shows that P_3 increases at a rate of 0.48 per unit increase in P_2.

This dose–response function applies when there is continuous exposure to a single toxicant. There are, however, other types of exposure that require a modified function. For example, individuals often are exposed to noncontinuous concentrations including pulsed exposures such as atmospheric or aquatic plumes released from a point source (e.g., Dixon 1977, Kendall et al. 2001). In these cases, an individual may be exposed to the toxicant only when it is located within the plume. An example of a dose–response function for pulsed exposures was developed by Giesy et al. (1999) to predict the mortality in *Daphnia* exposed to runoff of chlorpyrifos into streams:

$$m(x,C(t)) = \frac{P_1}{1 + e^{(2.2/P_3)(P_2 - \sum_{t=bd}^{\tau} C(t))}} \tag{5.5}$$

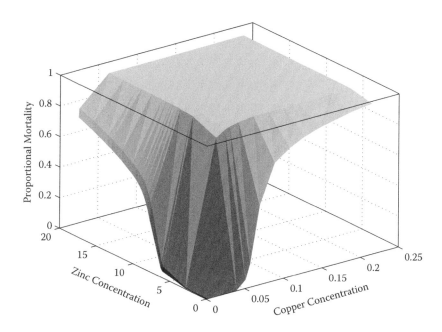

FIGURE 5.6 Dose–response model of tadpole mortality following exposure to separate metal concentrations and mixtures of Cu and Zn. (Data from Jennifer Gottschalk, "Copper and Zinc Toxicity to the Gray Treefrog (*Hyla chrysocelis*) and the Northern Leopard Frog (*Rana pipiens*)," Master of Science thesis, Clemson University, 1995.)

where

$m(x, C(t))$ = cumulative mortality of cohort x to $C(t)$ from birth date (*bd*) to time tau
P_1 = maximum cumulative mortality
P_2 = cumulative concentration yielding peak mortality rate
P_3 = cumulative concentration between 0.1 and 0.5 total mortality
$C(t)$ = chlorpyrifos concentration at time t

Instead of continuous exposure to a single toxicant, Q, as in Equation (5.4), the animals are exposed to the cumulative concentration of a toxicant over a "window" of time,

$$\sum_{t=bd}^{\tau} C(t)$$

Another type of dose response is that resulting from exposure to a mixture of toxicants. For example, Gottschalk (1995) found that exposure to mixtures of copper and zinc significantly increased mortality in *Rana pipiens* tadpoles (Figure 5.6).

In this figure, the dose responses to the separate metals are on the edges of the response surface above their respective axes. Mortality from a 96-hour exposure to copper alone reaches an asymptote of 0.97 at about 0.25mg/l; the response to zinc alone approaches an asymptote around 0.80 at 20mg/l. The dose response to mixtures, however, shows a synergistic effect: mortality reaches 99–100% at mixtures of copper and zinc at or above 0.04mg/l of copper and 6.2mg/l of zinc. The dose–response model was obtained using Equation (5.4) in which the parameters P_1, P_2, and P_3 were written as functions of copper and zinc:

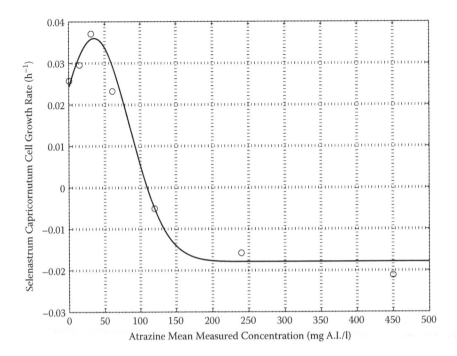

FIGURE 5.7   Graded response of *S. capricornutum* growth rate to atrazine concentration.

$$P_1 = (b_5 - b_4)(1 - e^{-(b_1 C + b_2 Z + b_3 CZ)}) + b_4$$

$$P_2 = (b_5 - b_4)e^{-(b_1 C + b_2 Z + b_3 CZ)} + b_4 \qquad\qquad (5.6)$$

$$P_3 = (b_5 - b_4)e^{-(b_1 C + b_2 Z + b_3 CZ)} + b_4$$

Note that the argument in the exponential function contains terms for each individual metal and an interaction term for mixtures $(+b_3 CZ)$. It follows that a dose response to mixtures showing an antagonistic effect would have a negative interaction term $(-b_3 CZ)$.

5.1.1.2   Graded Response

A graded response can be described by a functional relation between the dose and the continuous response that either increases or decreases as the dose increases. An example of such a response is the effect of atrazine concentration on the growth rate of the alga, *Selenastrum capricornutum* (Hoberg 1993) (Figure 5.7).

This response also illustrates the phenomenon of hysteresis, where a small increase in the dose results in a beneficial response—in this case, an increase in cell growth. After the hysteresis effect, the response changes to a more typical negative response with increasing the dose. The example of cell growth in *S. capricornutum* shows that with increasing the dose there is a negative response of decreasing cell growth with increasing the dose.

5.1.2   Toxicokinetics

As we have observed, the dose depends upon the concentration of the chemical at the site and the duration of the exposure. To estimate the concentration at the site we need to know how much of the

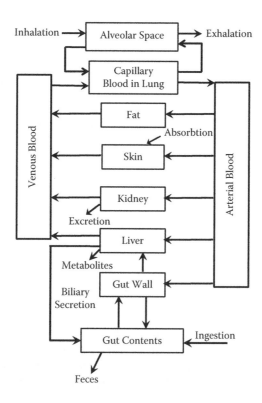

FIGURE 5.8 Example of a physiologically based toxicokinetic (PBTK) model.

chemical is taken up and absorbed by the organism, where the chemical is distributed among the organism's tissues and organs, and the rate at which the chemical is excreted from the same tissues and organs.

The dynamics of the disposition of a toxicant in the body of an organism is known as *toxicokinetics*. This involves the changes over time of the concentration of a toxicant in various tissues and the rate processes that control the movement from one part of the organism to another. These processes include uptake, absorption, distribution, and elimination. An information flow diagram (Figure 5.8) illustrates the dynamic processes in toxicokinetics. In Example 5.3 we describe a conceptual physiologically based toxicokinetic (PBTK) model of dieldrin toxicokinetics in American badgers (*Taxidea taxus*). The case study in Chapter 12 describes a PBTK model in greater detail. Even greater details on toxicokinetic modeling can be found in Reddy et al. (2005).

5.1.3 PHYSIOLOGICAL PROCESSES

5.1.3.1 Uptake

The two main routes by which toxicants enter the body are through the lungs (or gills) and gastrointestinal tract (Figure 5.8).

The rate of uptake of a toxic gas or vapor by the lungs, dU_L/dt (mg·min^{-1}), can be expressed by the differential equation:

$$\frac{dU_L}{dt} = 10^{-6} Y \cdot V_T \cdot f \tag{5.7}$$

where

Y = exposure concentration (mg·m^{-3}),
V_T = tidal volume (ml·breath^{-1}), and
f = breathing frequency (breaths·minute^{-1})
10^{-6} = constant to convert m^3 to ml

Uptake by gills has been modeled using Fick's diffusion equation (Nichols et al. 1990, Albers and Dixon 2002).

$$F^G = k_X^G \left(f_W \cdot C_W^{aff,G} - f_B \cdot C_B^{aff,G} \right)$$

(5.8)

where

F^G = flux of perchlorate across the gills, mg·kg^{-1}·h^{-1}
k_X^G = exchange coefficient
f_W = ratio of free chemical in exposure water to total concentration
f_B = ratio of free to total perchlorate in blood
$C_W^{aff,G}$ = total concentration of perchlorate in exposure water, mg·kg^{-1}
$C_B^{aff,G}$ = concentration of perchlorate in the blood afferent to the gills, mg·kg^{-1}

See Section 5.1.3.2 for a more complete discussion of Fick's law.

Uptake by ingestion is a function of toxicant concentration in the ingested material and feeding rate. For example, ingestion of insecticide granules has been modeled by Solomon et al. (2001) and used in the case study in Chapter 10. The rate of ingestion G_t, is described by Equation (5.9):

$$G_t = \frac{p \cdot Wg \cdot Vg \cdot Dg \cdot 1000 \cdot Ng_t}{Wt}$$

(5.9)

where

p = proportion of time spent in granule-treated areas
Wg = weight of granules, mg
Vg = granule insecticide concentration, mg·mg^{-1}
Wt = consumer body weight, g
Dg = dissipation rate of granular insecticide, h^{-1}
Ng_t = number of granules ingested

The 1000 in the numerator makes the units on the right-hand side (RHS) of the equation conform to those on the left-hand side (LHS): mg·kg^{-1}·h^{-1}

Example 5.1

This example solves the inhalation model (Equation [5.7]) with the addition of an elimination term. The m-file **resp3** contains the function Up, which defines the differential equation. The m-file **resp4** contains the initial conditions, the differential equation (d.e.) solver, and the plot statements.

```
%resp3
function Up=resp(t,U);
Y=3.0;       %Exposure concentration,mg/m^3)
Vt=22;       %Tidal volume, ml/breath
f=8.0;       %Breathing frequency, breaths/minute
c=.01;       %Elimination rate constant, 1/min
Up=(1e-6)*Y*Vt*f - c*U;
```

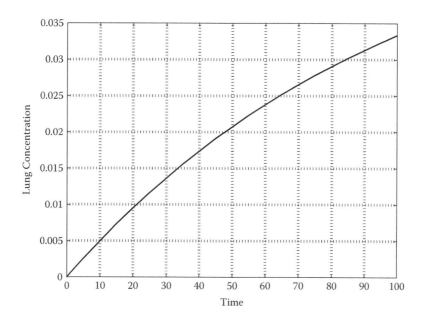

FIGURE 5.9   Plot of inhalation uptake over time.

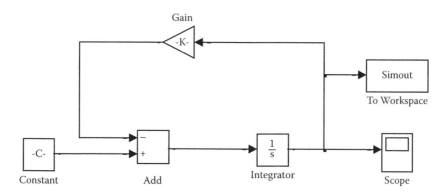

FIGURE 5.10   Simulink block diagram of inhalation uptake model **inhale**.

```
%resp4
U0=0;
tspan=[0.0 100.0];
[t,U]=ode23('resp3',tspan,U0);
plot(t,U);
xlabel ('Time');
ylabel('Lung Concentration');
grid;
```

The resulting graphical output is shown in Figure 5.9.

The same model can be expressed as a block diagram in Simulink (Figure 5.10). The modeled lung concentration data is output to two sink blocks: the **To Workspace** block with the label **simout** and the **Scope** block.

Clicking the **Scope** block creates a figure in the **Scope** window. To save the scope figure that looks identical to the **Scope** window, click the **Parameters** button (second from left) in the scope window. Then, in the **Data history** pane, check the **Save data to workspace** option

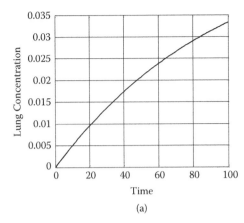
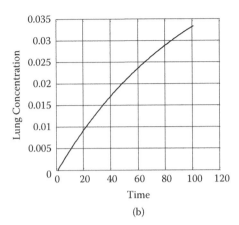

FIGURE 5.11   Plot of lung concentration over time from Simulink output (a) scope and (b) workspace.

and choose **Structure with time** from the **Format** list. To run the simulation, click on the **Simulation** button and then **Start** on the drop-down menu. The scope figure can be created with the command **simplot(ScopeData)** in the Command Window (Figure 5.11a). To plot the workspace figure, right click the **To Workspace** block and then **To Workspace Parameters**. Select the save format as an array. Then type plot **(simout)** in the Command Window to produce the graph. The plot is edited using the **Edit Plot** button in the figure window to add labels and increase font size and line width. Notice the similarity between Figures 5.9 and 5.11.

5.1.3.2 Absorption

Absorption is the process by which a toxicant enters the bloodstream, or in the case of plants, the phloem. The process usually involves the movement of the toxicant across cell membranes, either by passive diffusion or active transport. This is true whether the membrane is the stratified epithelium of the skin, the cell layers separating the lungs and the gastrointestinal tract from the bloodstream, or the bloodstream from the cells of the target organs. Active transport is the process of moving particles across a membrane against a concentration gradient.

Active transport differs from diffusion in that toxicants move against electrochemical or concentration gradients and the movement requires the expenditure of energy.

Most toxicants cross membranes by simple diffusion, which can be described by Fick's law:

$$\frac{\partial C(x,t)}{\partial t} = D\frac{\partial^2 C(x,t)}{\partial x^2} \tag{5.10}$$

According to this equation, the rate of change in the toxicant concentration C on one side of the membrane is proportional to the nonlinear change in concentration relative to distance x moved within the membrane. In dilute concentrations, the diffusion coefficient D can be assumed constant and has units (length)^2/time. If the concentrations on the surfaces of the membrane C_1, C_2 are maintained at constant levels, a steady state is reached in which the concentration changes linearly from C_1 to C_2. The rate of diffusion F of the toxicant is then the same throughout the membrane and can be expressed by:

$$F = \frac{\partial C}{\partial t} = -D\frac{\partial C}{\partial x} \cong D\frac{(C_1 - C_2)}{l} \tag{5.11}$$

where l is the membrane thickness.

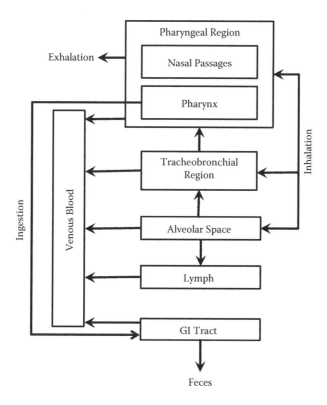

FIGURE 5.12 Material flow diagram for the absorption and transport of inhaled toxicants.

Gastrointestinal tract. Once taken up through ingestion, toxicants can be absorbed by the gastrointestinal (GI) tract. Lipid soluble organic acids or bases tend to be absorbed by simple diffusion. Active transport mechanisms are responsible for the absorption of some toxicants such as lead. Bioconversion or degradation of toxicants is also possible in the GI tract through activity of the digestive enzymes or intestinal flora.

Lungs. Absorption of toxicants, (gases, vapors, aerosols, and particles) occurs in the lungs following inhalation. Depending upon the type of toxicant, not all compartments of the inhalation pathway (Figure 5.12) will need to be included in a model.

The major processes affecting absorption involve the solubility of the toxicant and chemical reactions between the toxicant and lung tissues. Highly water-soluble gases may be absorbed through the mucous membranes in the nasal passages. For toxicants with little or no reactivity, the blood-to-gas partition coefficient (solubility ratio) of a gas or vapor determines the rate of absorption. The absorption rate of gases with a low solubility ratio depends upon the rate of removal from the lungs by blood flow. For gases with a high solubility ratio, the absorption rate is limited primarily by breathing rate, the removal rate of the gas in the capillaries is relative rapid, and the concentration in the capillaries will be approximately zero ($C_2 = 0$ in Equation [5.11]). Deposition of aerosols and particles greater than 5 μm occurs in the nasopharyngeal region. Insoluble particles deposited in the nasal region are moved to the pharynx by ciliary action until they are swallowed and enter the GI tract. A similar mechanism removes particles dissolved in the nasal mucus. Dissolved particles also may enter the bloodstream by absorption through the nasal epithelium.

Particles 2 to 5 μm are deposited in the trachea and bronchia of the lungs. From there they are removed by the ciliated mucous layer to the pharynx where they may be swallowed. Particles ≤1 μm are deposited in the alveoli of the lungs where they can be absorbed into the bloodstream or enveloped by macrophages and removed via the lymphatic system.

A more detailed discussion of modeling pulmonary gas exchange can be found in Rees et al. (2001).

Skin. The skin is only slightly permeable and absorption through it is relatively slow compared with other uptake pathways. When the skin is exposed to an exogenous toxicant, the rate of penetration through the skin is constant, after an initial period of gradual increase. Circulation beneath the skin's surface is sufficient to transport the toxicant away from the penetration site resulting in a dynamic equilibrium. Thus, the absorption rate is proportional to the difference in concentration on either side of the skin and can be described by Fick's Law (Scheuplein and Blank 1971, Schaefer and Jamoulle 1988).

Example 5.2

The m-files in this example simulate the absorption of a toxic compound by the skin. The model uses Equation (5.11) to simulate the amount of the compound on the external skin surface and the percutaneous concentration, assuming a constant initial exposure and transport of the compound away from the skin via the blood stream.

```
%function skin3
%model of skin absorption using Fick's Law
function qp=skin3(t,q);
k1=0.05;
k2=0.05;
dq1dt=-k1*(q(1)-q(2));          %skin surface concentration d.e.
dq2dt=k1*(q(1)-q(2)) - k2*q(2); %percutaneous concentration d.e.
qp=[dq1dt;dq2dt];


%m-file skin4
%program to simulate skin absorption
q0=[100 10]';
tspan=[0 100];
[t,q]=ode23('skin3',tspan,q0);
hold on
plot(t,q);
grid on;
xlabel('Hours');
ylabel('Concentration');
title('Skin Toxicokinetics');
```

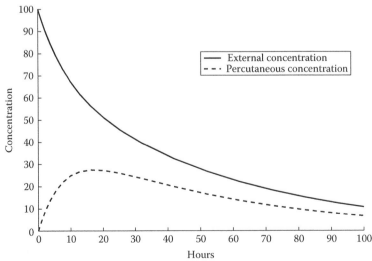

FIGURE 5.13   Uptake and removal of toxic compound by the skin.

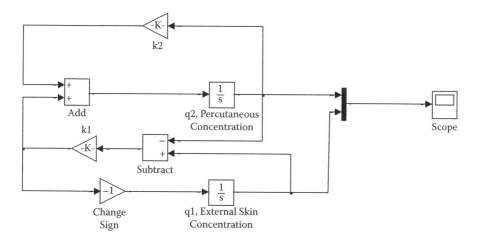

FIGURE 5.14 Simulink block diagram of skin absorption model.

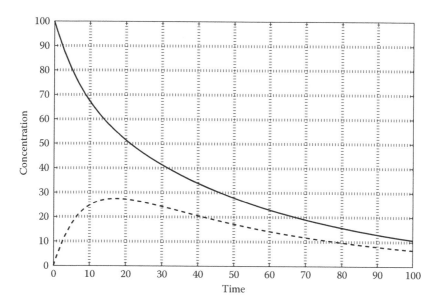

FIGURE 5.15 Plot of external and percutaneous skin concentrations from the Simulink skin absorption model.

The results in Figure 5.13 show an exponential decrease in the toxic compound on the skin surface from an initial value of 100. The percutaneous concentration increases from an initial value of 0 to a maximum of 30 before decreasing gradually toward zero after 100 hours.

In this example, we also examined the absorption model using Simulink. A block diagram of the model, titled **SKIN** is shown in Figure 5.14. To run the simulation, click the **Simulation** button in the **SKIN** model window and then click **Start** on the drop-down menu.

To plot the figure in the scope window, click the **Scope** block to create a figure in the **Scope** window. Save the scope figure by clicking the **Parameters** button (second from left in the scope window). Then, in the **Data history** pane, check the **Save data to workspace** option and choose **Structure with time** from the **Format** list. The scope figure can be created with the command **simplot(ScopeData)** in the Command Window (Figure 5.15). The figure then can be edited by clicking the **Edit Plot** icon and selecting the **Property Editor** in the **View** drop-down menu.

5.1.3.3 Distribution

Once a toxicant has entered the body, it is transported in the bloodstream to tissues and organs in the body (Figure 5.8). The rate of blood transport is relatively rapid. The penetration of a toxicant into the cells of a particular organ or tissue is determined by the rate of diffusion or active transport, which in turn depends upon the affinity of the toxicant for the particular tissue. Diffusion is the typical process for small water-soluble molecules and ions that pass through aqueous channels or pores in the cell membrane and lipid-soluble molecules that cross the cell membrane itself. Active transport is required for polar molecules and ions with a molecular weight ≤ 50.

Toxicants often show an affinity for a particular organ or tissue. The tissue or organ where a toxicant accumulates may be the site of toxic action but more often, the toxicant will accumulate in an inactive site. Storage in an inactive site reduces the plasma concentration and thus prevents the toxicant from reaching the target organ. Equilibrium of a toxicant's concentration exists between the storage site and blood plasma. As the concentration in plasma is decreased by excretion or bio-transformation, more of the toxicant is released from the storage site.

Several tissues or organs are storage sites for xenobiotics. Plasma proteins bind some toxicants such as the insecticide dieldrin. It is possible for one toxicant to displace a second chemical at the binding site on plasma proteins. The concentration of the freed toxicant in plasma then increases with the potential for increased transport and deposition in the target organ, resulting in a toxic reaction. Both the liver and kidney bind many different chemicals, including heavy metals. Accumulation in adipose tissue is typical of lipophilic compounds such as organochlorines and polychlorinated biphenyls (PCBs). The storage of toxicants in body fat can prevent a toxic effect in the target organ. Under periods of stress, however, body fat can be mobilized rapidly, resulting in a possible toxic dose to the target organ.

Example 5.3

This example illustrates a model, developed by Henriques (1996), of the exposure of American badgers (*Taxidea taxus*) to the pesticide dieldrin. It was developed to predict the uptake and distribution of dieldrin in badgers at the Rocky Mountain Arsenal (RMA) in Colorado. Exposure pathways included inhalation and adsorption from the soil surface and ingestion of both soil and prey items, mostly black-tailed prairie dogs (*Cynomys ludovicianus*) (Figure 5.16). Significant concentrations of dieldrin have been found in the soil at RMA. Because of their burrowing habits, dieldrin comes into contact with the skin of the badgers. Therefore, dermal absorption is a significant exposure pathway. Dieldrin particles deposited in the trachea and bronchia of the lungs are removed by the ciliated mucous layer and then swallowed where they enter the GI tract. Particles ≤1 μm are deposited in the alveoli of the lungs where they can be absorbed into the bloodstream (Section 5.1.3.2).

Ingested dieldrin, whether from soil or prey, passes through the GI tract, where a fraction is absorbed in the small intestine, transported through the hepatic portal vein to the liver, and then enters the bloodstream. Model partitioning coefficients are determined by the ratio of dieldrin in the tissue relative to that in the blood at equilibrium. Partitioning coefficients are greater in lipid tissues such as fat, liver, and brain. For example, the ratio of dieldrin concentration in fat to that in blood is >130:1 (Hayes 1974). Dieldrin has been shown to cross the placenta (Polishuk et al. 1977) and is found in mother's milk (ATSDR 2002). These exposure pathways, as well as dermal absorption through newly formed skin, make the risk to newborn badger pups particularly high. A more extensive description of a PBTK model is found in Chapter 12.

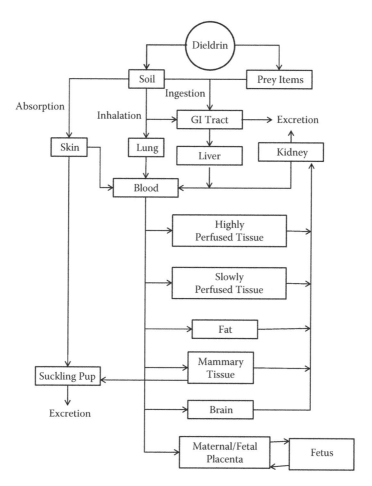

FIGURE 5.16 Badger PBTK model flow diagram. (Modified with permission from William Henriques, "A Model of Spatial and Temporal Exposure and Effect of Dieldrin on Badgers at the Rocky Mountain Arsenal," Ph.D. Diss., Clemson University, 1996.)

5.1.3.4 Excretion

There are several routes by which toxicants are removed from the body (Figure 5.8). Most toxicants are excreted by the kidneys via urine. Other routes, in order of importance, are biliary excretion via the feces and by the lung via expired air.

Toxicants are excreted in urine by several processes, including filtration at the glomeruli, and both passive diffusion and active transport by the tubules of the kidney. The relative importance of the processes depends upon the toxicant and the species being considered.

Fecal excretion is a complex process involving several different pathways. Not all of an ingested toxic substance is absorbed in the gastrointestinal tract and therefore is excreted directly. The fraction of a toxicant that is absorbed by the gastrointestinal tract is transported by the blood to the liver where a large percentage of the absorbed compound may be extracted before the blood reaches other parts of the body.

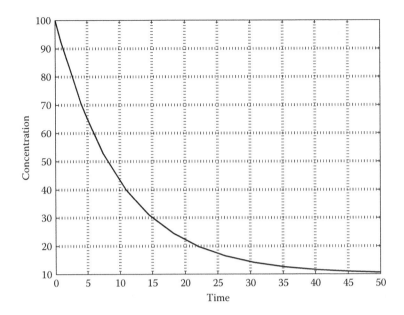

FIGURE 5.17   Amount of toxic substance remaining in compartment during constant excretion.

Example 5.4

In Example 2.2, we described a simple model of excretion. In this example, we describe the m-files to solve the excretion model. The m-file **expdecay** contains the d.e. and model parameters, p2 and p4. The m-file **expdecay2** contains the initial condition, q0, the simulation time, the d.e. solver, and the plot statements.

```
% m-file expdecay
function qdot = expdecay(t,q);
p2=0.1;     %excretion rate constant, 1/t
p4=10.;     %minimum (background) compartment concentration
qdot = -p2*(q-p4);


%m-file expdecay2
q0=100;
tspan=[0 50];
[t,q]=ode23('expdecay',tspan,q0);
plot (t,q)
grid;
xlabel('Time');
ylabel('Concentration');
```

Results of the simulation in Figure 5.17 show an exponential decrease in compartment concentration from an initial value of 100 to an asymptotic value of 10 (p4).

5.1.4   Biological Processes

5.1.4.1   Reproduction

Toxicants can adversely affect reproduction in several ways. Environmental estrogens, such as dichlorodiphenyltrichloroethane (DDT) and PCBs, may cause structural changes in the prenatal reproductive organs, causing female characteristics to persist in males, and skewing the sex ratio

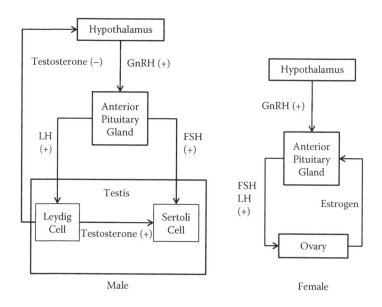

FIGURE 5.18 Simplified material flow diagram of the mammalian reproductive cycle in males (left) and females (right).

in the population toward females. Kraft mill effluent has been shown to cause masculinization in fish living in the streams receiving the effluent. In mammals, toxicants can prevent the implantation of the fertilized egg. In birds, organochlorines are well known for their eggshell-thinning effect, which greatly reduces egg survival. Toxicant effects may cause stress-related reduction in egg and sperm production. Behavior also can be affected by toxicants resulting in reduced care of eggs or offspring. All of these effects lead to reduced probability of successful reproduction in an exposed individual and a lowered population birth rate.

Reproduction can be modeled at the hormonal level and the individual level. At the hormonal level, the reproductive process is regulated by the hypothalamic-pituitary-gonadal axis (Figure 5.18).

A wide variety of toxicants can interfere with the reproductive hormonal system. Some triazine herbicides such as atrazine have been shown to interfere with the release of the gonadotropin-releasing hormone (GnRH) by the hypothalamus and subsequent release of the luteinizing hormone by the pituitary in rats (Cooper et al. 1996, 2000).

At the individual level, the effects of toxicants on reproductive rates have been observed whether the exact causal mechanism is understood. Effects of toxicants on the development of a fetus may lead to its death before or after birth. If mortality occurs before birth, it can be considered a failure to reproduce successfully. If it occurs after parturition, it can be considered as the death of the individual organism. Toxicants not only can have a teratogenic effect on the fetus, but also can adversely affect that organism's ability to reproduce when it reaches maturity. In vertebrates, an individual female can have a discrete number of young according to some probability mass function. That is, there is a probability associated with having zero, one, two, etc., young (see Example 4.3).

Example 5.5

This example selects the number of young produced in a single litter and generates a histogram of the number in each litter size for a population of 1000 females. The probability of a certain number of young in a litter follows the empirical probability mass function in Example 4.3. The probability of zero young is 0.15, one = 0.30, two = 0.40, and three = 0.15. The MATLAB m-file to generate the random variates is:

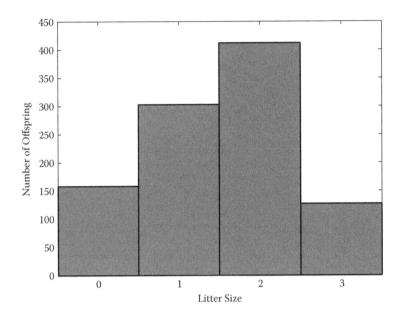

FIGURE 5.19   Histogram of the number of young in litter sizes 0–3 out of 1,000 litters produced.

```
% m-file repro3
% program to generate numbers of offspring
% according to an empirical PMF

for i=1:1000
    uu=rand;
    if uu <= 0.15;
       egg(i)=0;
    elseif uu> 0.15 && uu <=0.45;
       egg(i)=1;
    elseif uu> 0.45 && uu<=0.85;
       egg(i)=2;
    elseif uu>0.85 && uu<=1.0;
       egg(i)=3;
    end;
end;

bins=[0 1 2 3];
hist(egg,bins)
xlabel('Litter Size')
ylabel('Number of Offspring')
```

The resulting histogram is shown in Figure 5.19.

5.1.4.2 Growth

Growth encompasses several different processes, including an increase in body size and the progression in the stage of development. For example, certain endocrine-disrupting compounds can delay metamorphosis in amphibians (Lefcort et al. 1998, Carr and Theodorakis 2006). A Markov chain can be used to model life-stage transition. As an example, we assume there are three life stages, e.g., egg, larvae, and adult, although more can be added to the model. Transitions from a stage occur at each time step, although additional stages can be added to allow for members of

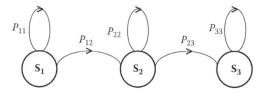

FIGURE 5.20 State-transition diagram for a life-stage transition Markov process.

a stage to remain in that stage for a longer period, that is, a single stage can be represented by several stages.

The state-transition diagram for this system is shown in Figure 5.20. States S_1 and S_2 are considered transition states in that once there is a transition out of those states, they cannot return to the previous state. State S_3 is an absorbing state in that once there is a transition to that state, there can be no transition out of that state.

Body size may be important to include in an individual-based model because many other processes are a function of body weight, such as respiration rate and ingestion rate. The factors affecting increase in body size are quite complex, especially at the genetic and hormonal level. The model of Weiss and Kavanau (1957) incorporates a relatively simple structure but with biologically relevant parameters. This (or a similar) model can be used to mimic body growth in an individual organism. The model also can be used to mimic separate tissues and organs in a PBTK model. Example 5.6 demonstrates a life-stage transition model and Example 5.7 the Weiss model of body growth.

Example 5.6

In this example, we use a Markov chain model of delay in life-stage transition resulting from exposure to a toxicant. All members of the population begin in the first stage. The transition probability matrix for the unexposed population is:

$$[p] = \begin{bmatrix} p_{11}=0.1 & p_{12}=1.-p_{11} & p_{13}=0.0 \\ p_{21}=0.0 & p_{22}=0.1 & p_{23}=1-p_{22} \\ p_{31}=0.0 & p_{32}=0.0 & p_{33}=1.0 \end{bmatrix} \qquad (5.12)$$

The probabilities of remaining in states 1 and 2 are 0.1. The probabilities of progressing to the next state for states 1 and 2 are 1 minus the probability of remaining in that state, or 0.90. The probability of remaining in state 3 is 1.0.

We used a hypothetical relation for the effect of a toxicant on transition probabilities. We assumed a constant exposure over time and the probability of an individual remaining in its current state was directly related to time of exposure (Figure 5.21).

These new transition probabilities yield the following matrix:

$$[p] = \begin{bmatrix} p_{11}=\dfrac{t}{t_{max}} & p_{12}=1.-p_{11} & p_{13}=0.0 \\ p_{21}=0.0 & p_{22}=\dfrac{t}{t_{max}} & p_{23}=1-p_{22} \\ p_{31}=0.0 & p_{32}=0.0 & p_{33}=1.0 \end{bmatrix} \qquad (5.13)$$

where t is time and t_{max} is the duration of exposure. The m-file to simulate exposed and nonexposed populations is as follows:

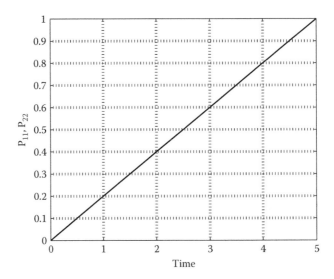

FIGURE 5.21   Hypothetical relation between the probability of remaining in the current state as a function of time (exposure).

```
% m-file stagegrowth
% Markov chain model of life stage growth
clear all

N=5;
%initial state values.  All individuals start in S1
S1(1)=100;S2(1)=0;S3(1)=0;

for i=1:N
    % transition probabilities without toxicant exposure
    %p11=0.1;p12=1.-p11;p13=0.0;
    %p21=0.0;p22=0.1;p23=1-p22;
    %p31=0.0;p32=0.0;p33=1.0;
    % transition probabilities with toxicant exposure
    p11=i./N;
    p12=1.-p11;p13=0.0;
    p21=0.0;
    p22=i./N;
    p23=1-p22;
    p31=0.0;p32=0.0;p33=1.0;
    % state equations
    S1(i+1)=p11*S1(i)+p21*S2(i)+p31*S3(i);
    S2(i+1)=p12*S1(i)+p22*S2(i)+p32*S3(i);
    S3(i+1)=p13*S1(i)+p23*S2(i)+p33*S3(i);
end
x=0:5;
plot(x,S1)
hold on
plot(x,S2,'r--')
plot(x,S3,'g-.')
grid on
xlabel('Time')
ylabel('Number of Individuals')
legend('Stage 1','Stage 2', 'Stage 3','location','North')
```

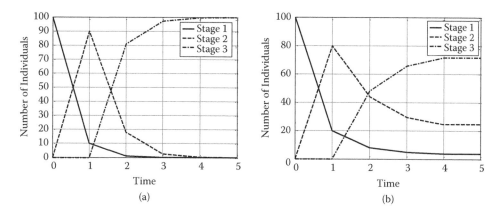

FIGURE 5.22 Transitions from different life stages, (a) transitions with constant probabilities, (b) transitions with probabilities as a function of exposure.

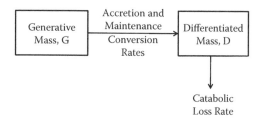

FIGURE 5.23 Flow diagram for body growth model. (Adapted from Paul Weiss and J. Lee Kavanau, "A Model of Growth and Growth Control in Mathematical Terms." *Journal of General Physiology* 41 (1957): 1–47. With permission from Rockefeller University.)

Note that the transition probabilities for the nonexposed population have been commented out using %. To run the simulation for nonexposed individuals, the transition probabilities for exposed individuals must be commented out, using %, and those % for the nonexposed probabilities removed. The results of the simulations are shown in Figure 5.22.

In both exposed and nonexposed individuals, the simulations start with all 100 individuals in the egg stage (stage 1). Without exposure, 90% of individuals leave the egg stage and enter the larval stage in the first time step. Exposed individuals have 80% of individuals transitioning from eggs to larvae in the same time. Nonexposed larvae also transition rapidly from larvae to adults with 80% of individuals reaching the adult stage in the second time step and 100% by the fifth time step. In the exposed population, only 50% reach adulthood by the end of step two and only a little more than 70% ever become adults. About 5% remain in the egg stage and 25% in the larvae stage.

The simplified Weiss and Kavanau (1957) model of body growth divides body mass into two compartments: generative mass, which is converted to differentiated mass (Figure 5.23).

The model is defined by a set of two differential equations, one for generative mass G and one for differentiated mass D:

$$\frac{dG}{dt} = (G \log 2)\left[1 - \frac{b(G^n - G_0^n)}{G_e^n - G_0^n}\right] - k_1 G\left[1 - \frac{(G^n - G_0^n)}{G_e^n - G_0^n}\right] - k_2 G$$

$$\frac{dD}{dt} = k_1 G\left[1 - \frac{(G^n - G_0^n)}{G_e^n - G_0^n}\right] + k_2 G - k_3 D$$

(5.14)

In the equation for G, the first term defines the rate of G formed. The basic rate of generative mass production is $G \log 2$. This rate is modified by the feedback multiplier,

$$\frac{b(G^n - G_0^n)}{G_e^n - G_0^n}$$

which reduces the basic rate according to the difference between G and the initial mass, G_0. The second term in the equation for G is the rate of conversion of generative mass to differentiated mass, which is also modified by a feedback multiplier. The third term differentiates a fraction of G for the maintenance of differentiated mass. The latter two terms defining the rate of differentiation are inputs in the equation for D. The last term in the D equation, $k_3 D$, is the rate of catabolic loss. The individual parameters in the model are:

b = ratio between feedback at equilibrium and complete inhibition,
k_1 = rate constant for conversion of G to D,
k_2 = rate constant for maintenance of D,
k_3 = rate constant for the catabolic loss of D,
G_0 = initial G mass at birth,
G_e = maximum adult G mass,
n = constant.

Example 5.7

This example uses the growth model from Weiss and Kavanau (1957) described previously. The simulation is for a chicken, with parameter values also from Weiss and Kavanau 1957, 23):

k_1 = 0.5077
k_2 = 0.1154
k_3 = 0.0089
b = 0.8335
G_0 = 0.02418
G_e = 4,096 G_0
D_e = 53,190 G_0
n = 0.5

The m-files for the model are a function (**growth**) containing the parameters and differential equations:

```
% function growth
% growth model from Weiss and Kavanau 1957
function Tdot = growth(t,T)
k1=0.5077; k2=0.1154; k3=0.0089;
Go=0.02418;             %initial generative mass at birth
Ge=4096*Go; De=53190*Go;
b=0.8335;
n=0.5;

term1=1-(b*(T(1)^n-Go^n)./(Ge^n-Go^n));
Gdot=(T(1)*log(2))*term1-k1*T(1)*term1-k2*T(1);
Ddot=k1*T(1)*term1+k2*T(1)-k3*T(2);
Mdot=(T(1)*log(2))*term1-k3*T(2);
Tdot=[Gdot; Ddot; Mdot];
```

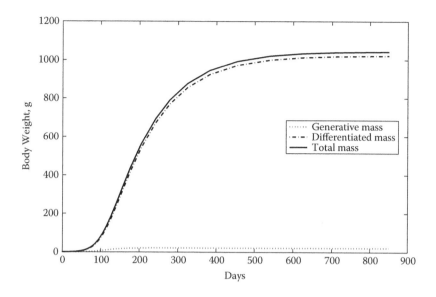

FIGURE 5.24 Predicted generative, differentiated, and total body mass of a chicken. (Adapted from Paul Weiss and J. Lee Kavanau, "A Model of Growth and Growth Control in Mathematical Terms." *Journal of General Physiology* 41 (1957):1–47. With permission from Rockefeller University.)

and a program to solve the equations and plot the results (**growth2**):

```
% m-file growth2
% program to solve growth equations
T0=[0.02418 0.0 0.02418]';
tspan=[0 850];

[t,T]=ode23('growth',tspan,T0);
plot(t,T);
legend('Generative Mass','Differentiated Mass',...
       'Total Mass','Location','East')
xlabel('Days')
ylabel('Body Weight, g')
```

The results of the simulation are shown in Figure 5.24. The graph shows a typical sigmoid growth curve. The rate of conversion from generative mass to differentiated mass is high enough to keep the generative mass at a small fraction of the total mass.

5.1.4.3 Death

Death of an organism can be caused by acute exposure to a toxic compound or by chronic exposure over a long period. Exposed individuals also can recover if the toxic compound is removed or the animal moves away from the compound. The health status of an organism then can be considered a system state variable. This leads to the question, "What is the nature of death?" Stacy (1969) defines it as the catastrophic failure of the organism system. As the organism is composed of many subsystems, the failure of any of these subsystems can lead to death.

In the case of acute exposure, the dose–response function described in Section 5.1.1.1, where the response is death, usually is an adequate death model. The steps to estimate the probability of an individual dying are: (1) estimate the exposure or dose concentration, (2) determine the probability of dying from the dose–response model, (3) obtain a sample random number from the uniform probability density function, and (4) determine whether the mortality probability exceeds the uniform

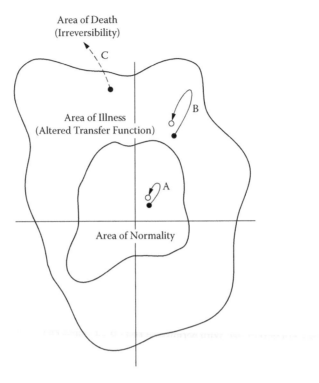

FIGURE 5.25   Diagram of the health status of individuals relative to two stressor variables represented by the x and y axes. Area A represents response to the stressors when the individual is in normality space, area B when the individual is in the space of abnormality but viability, and area C when the individual is in the area of death. (From R. W. Stacy, "The Comprehensive Patient-Monitoring Concept." In *Computers in Biomedical Research, Vol. III*, 253–276 (Copyright 1969). With permission from Elsevier.)

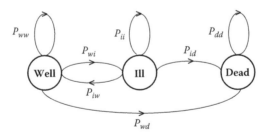

FIGURE 5.26   State-transition diagram for individual health status.

random variate, in which case the individual dies. An example of this procedure is illustrated in the case study in Chapter 10.

When there is chronic exposure, and there is a more gradual change in an organism's health status, these health states can be shown graphically (Figure 5.25, from Stacy 1969). In this figure, we assume there are three states of health: normality, illness, and death. The two-dimensional plot shows the parameter space of each health state as a function of two parameters, one on the x axis and one on the y axis. The same concept can be extended to multidimensional parameter space.

Death, following a sequence of changes in an individual's state of health, has been modeled as a Markov process (Beck and Paulker 1983, Sonnenberg and Beck 1993). In this Markov process model, an individual is in one of three states: well, ill, or dead (Figure 5.26). Transition probabilities define the probability of moving among states. Obviously, no one can leave the dead state. Thus, the probability of remaining in the dead state, P_{dd}, is 1.0. This type of state is referred to as an *absorbing state*.

Example 5.8

In this example, we exercise the Markov state-transition model, with transitions from wellness to illness to death, described previously. The model is from Sonnenberg and Beck (1993); transition probabilities are from their Table 1. We simulate the Markov process, first with no exposure to a toxicant, followed by a simulation with exposure. Instead of altering the probabilities for exposed and nonexposed individuals in the program, we model each population separately and then plot the results of each simulation in a single graph. The m-file for this model is as follows:

```
% m-file deathmarkov.m
% Markov chain model of death
% n = nonexposed, y = exposed
clear all

N=14;
%initial state values.  All individuals start in S1
S1n=zeros(1,N+1);
S2n=zeros(1,N+1);
S3n=zeros(1,N+1);
S1y=zeros(1,N+1);
S2y=zeros(1,N+1);
S3y=zeros(1,N+1);
S1n(1)=100;S2n(1)=0;S3n(1)=0;
S1y(1)=100;S2y(1)=0;S3y(1)=0;

for i=1:N
    % transition probabilities without exposure
    p11n=0.6;p12n=0.2;p13n=1-p11n-p12n;
    p21n=0.0;p22n=0.6;p23n=1-p21n-p22n;
    p31n=0.0;p32n=0.0;p33n=1.0;
    % transition probabilities with exposure
    p11y=0.2;p12y=0.4;p13y=1-p11y-p12y;
    p21y=0.0;p22y=0.4;p23y=1-p21y-p22y;
    p31y=0.0;p32y=0.0;p33y=1.0;

    % state equations
    S1n(i+1)=p11n*S1n(i)+p21n*S2n(i)+p31n*S3n(i);
    S2n(i+1)=p12n*S1n(i)+p22n*S2n(i)+p32n*S3n(i);
    S3n(i+1)=p13n*S1n(i)+p23n*S2n(i)+p33n*S3n(i);
    S1y(i+1)=p11y*S1y(i)+p21y*S2y(i)+p31y*S3y(i);
    S2y(i+1)=p12y*S1y(i)+p22y*S2y(i)+p32y*S3y(i);
    S3y(i+1)=p13y*S1y(i)+p23y*S2y(i)+p33y*S3y(i);
end
x=0:N;
%plot nonexposed
subplot(2,1,1)
plot(x,S1n,'b-',x,S2n,'r--',x,S3n,'g-.','LineWidth',2)
xlabel('Time')
ylabel('Number of Individuals')
title('Non-Exposed')
legend('Well','Ill', 'Dead')
grid on
```

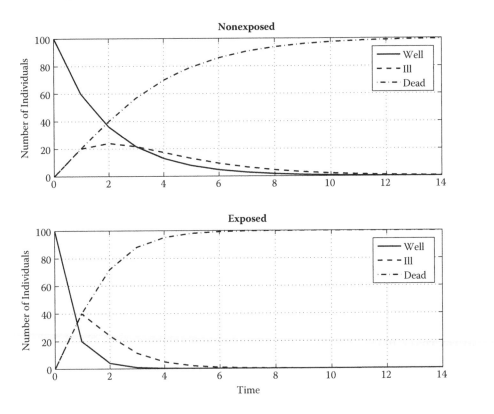

FIGURE 5.27 Results of Markov model of health status, with and without exposure to a toxicant. (From Sonnenberg, F. A., and J. R. Beck. 1993. "Markov Models in Medical Decision Making: A Practical Guide." *Medical Decision Making* 13:322–338. With permission from Sage Publishing.)

```
%plot exposed
subplot(2,1,2)
plot(x,S1y,'b-',x,S2y,'r--',x,S3y,'g-.','LineWidth',2)
%hold all

grid on
xlabel('Time')
ylabel('Number of Individuals')
title('Exposed')
legend('Well','Ill', 'Dead')
```

In the exposed population, we doubled the probability of going from well to ill from 0.2 to 0.4 and increased the probability of going from ill to death from 0.4 to 0.6. The results (Figure 5.27) show a much more rapid transition from well to ill, with about 70% remaining well after one time step for the nonexposed population and about 20% for the exposed population. This caused a temporary increase in the ill population after one time step (40% for the exposed population compared to 20% for the nonexposed population). The increase in the transition from ill to dead reduced the number of individuals in the ill population and increased the number in the dead population. The time for the entire population to die also was greatly reduced in the exposed population.

5.1.4.4 Movement

Sheeler-Gordon and Dixon (2000) described three types of models of animal movement: (1) matching spatial patterns of observed behavior (Siniff and Jessen 1969), (2) rules based upon mechanisms governing the response of an individual to its environment (Wolff 1994), or (3) theoretical constructs such as random walk models (Holgate 1971, Tyler and Rose 1994). Models that rely on pattern matching require significant amounts of data collected over a period long enough to define the probability distributions of direction and distance moved. These models require little information on the biology or behavior of the animal. Models that are only pattern based are limited to situations for which the data are available and cannot be extended to different habitats. Mechanistic rule-based movement models have some advantages over pattern-matching models. They can include variables characterizing the environment to which an animal will respond, such as food availability. This allows the model to predict unusual situations such as exposure to environmental contaminants. Information on an animal's relationship to its habitat also can be included in the model. Stochastic process models include random walk models and constrained random walk models. These models use transition probabilities that define the direction and distance moved, based on the animal's position relative to an activity center (Dunn and Gipson 1977, Dunn 1978).

Different approaches to movement modeling depend upon the distance an animal moves, which is often determined by the animal's feeding behavior (herbivore, omnivore, or predator). Long-distance movement can be modeled as a probability distribution of distances and/or directions (e.g., Allen and Ernest 2002). Short-distance movement can be based upon daily activity within a home range or travel over a larger area in a given time interval. One approach to modeling short-distance movement is a cellular automaton model.

A cellular automaton (CA) can be either continuous or discrete, but we limit our example to discrete cellular automata. We can describe an animal's area of movement as a two-dimensional grid divided into a finite number of cells. The neighborhood of a given cell is defined by the cell and its surrounding eight cells. The initial conditions for the model are the number of individual animals in each cell at the beginning of the simulation. At each time step, a rule specifies the probability of an animal's movement. For example, each animal could be restricted either to staying in its present cell or moving to another cell in its neighborhood. A rule also is needed for an animal that moves off the grid. One such rule is to allow a new animal to enter the grid on the edge opposite the edge where the first animal left.

A CA model can be implemented in MATLAB as a multidimensional structure array, which is an extension of a rectangular structure array. For example, the following statements define attributes of three raccoons, including their location:

```
raccoon(1).age = 1;
raccoon(1).sex = 'M';
raccoon(1).location = [3 8];
raccoon(1).weight = 15;
raccoon(1).perchlorate = 0.38;
raccoon(2).age = 3;
raccoon(2).sex = 'F';
raccoon(2).location = [6 6];
raccoon(2).weight = 17;
raccoon(2).perchlorate = 0.45;
raccoon(3).age = 2;
raccoon(3).sex = 'F';
raccoon(3).location = [5 3];
raccoon(3).weight = 13;
raccoon(3).perchlorate = 0.13;
```

An example of a multidimensional structure array for the raccoon model is shown in Figure 5.28.

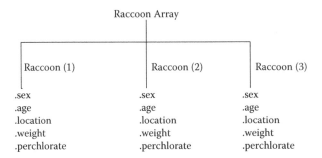

FIGURE 5.28   Diagram of the raccoon multidimensional structure array.

Example 5.9

This example is based upon a CA model developed by Sheeler (2002) to predict the spread of rabies in raccoons. The initial step is to distribute the population of raccoons on the grid. In the example, we randomly assigned five raccoons to a 10 × 10 square grid. As raccoons are nocturnal, we limited hours of movement between 0000 and 0600 hours and 1900 and 2400 hours. The MATLAB m-file to generate the movement is **movement.m** on the CD. The resulting 12-hour movement is shown in Figure 5.29.

5.1.4.5 Homeostasis

One effect that is particularly complex from a modeling perspective is the disruption of homeostasis.

Definition 5.1: *Homeostasis* is the tendency to maintain normal, internal stability in an organism by coordinated responses of the organ systems that automatically compensate for environmental changes.   ∎

Homeostasis is found in many subsystems in the body, including those that regulate minerals such as calcium and the metals, such as zinc. Hormonal systems are a good example of the tendency of systems to maintain equilibrium. An example is the thyroid hormonal regulatory system. In this system, thyrotropin-releasing hormone (TRH), which is produced by the hypothalamus, regulates the production of the thyroid-stimulating hormone, TSH, by the pituitary gland. TSH, in turn, stimulates the production of the two thyroid hormones, T3 and T4. The levels of T3 and T4 in the blood form a negative feedback loop to affect the release of TSH by the pituitary gland (Figure 5.30). When the levels of T_3 and T_4 are low, the production of TSH is increased, and conversely, when levels of T_3 and T_4 are high, production of TSH is decreased.

The model of the above system was adapted from Saratchandran et al. (1976). It is a simplified model of the thyroid system and is presented here only to illustrate the homeostasis of the system. The change in the plasma concentration of TSH is written as a function of TRH secretion rate:

$$\frac{d(TSH)}{dt} = aHT \cdot \log TRH(t) - k_3 \cdot FT3P(t) - k_5\left[FT3P(t) - FT3P'\right] - k_L \cdot TSH(t) \qquad (5.15)$$

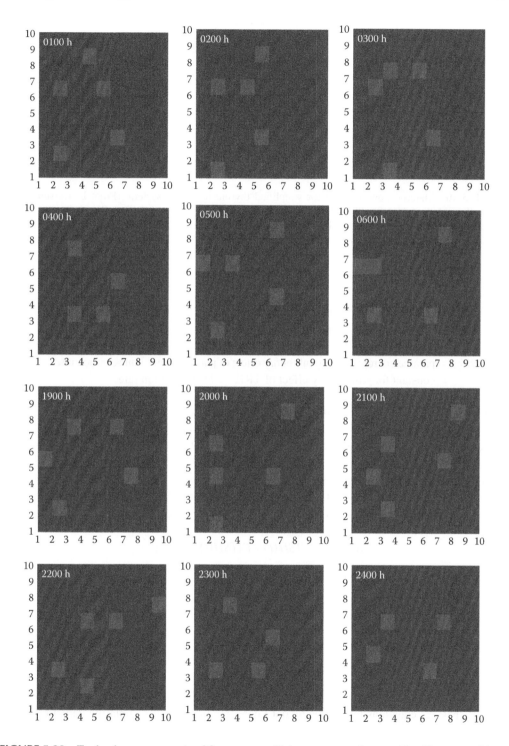

FIGURE 5.29 Twelve-hour movements of five raccoons (light gray squares) over a 10 × 10 square grid.

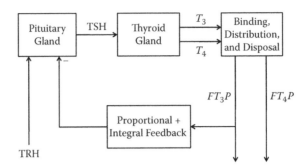

FIGURE 5.30 Material flow diagram for simplified thyroid hormonal system. (From E. Carson, et al., In *Modelling Methodology for Physiology and Medicine*, ed. E. Carson and C. Cobelli, 15–44. New York: Academic Press, 2001, 15–44, after P. Saratchandran et al., "An Improved Mathematical Model of the Human Thyroid Hormone Regulation." *Clinical Endocrinology* 5 (1976):473–483. With permission John Wiley and Sons.)

where

$TSH(t)$ = concentration of TSH in plasma at time t

$TRH(t)$ = secretion rate of TRH in plasma at time t

$FT3P(t)$ = concentration of free fraction of T3 in plasma at time t

k_3 and k_5 = parameters representing inhibitory effect of plasma-free T3 on secretion rate of TSH

aHT = constant for stimulatory effect of TRH on TSH secretion rate

k_L = loss rate of TSH in plasma

$FT3P'$ = represents the normal concentration of free fraction of T3 in plasma

The controlling part of Equation (5.15) is the term $k_5[FT3P-FT3P']$. When the concentration of plasma-free T3 is greater than the normal concentration, the term is negative resulting in a decrease in TSH. When the difference between $FT3P$ and $FT3P'$ is negative, the term is positive and the TSH concentration increases until equilibrium is restored. The rates of secretion of plasma T3 and T4 also have a controlling term that is a function of the TSH concentration.

$$SRT3 = k_1 \cdot TSH(t) + K[TSH(t) - TSH']$$

$$SRT4 = k_2 \cdot TSH(t) + K[TSH(t) - TSH']$$

(5.16)

where

$SRT3$ = secretion rate of T3

$SRT4$ = secretion rate of T4

TSH' = normal value of TSH in plasma

k_1, k_2, and K are parameters representing the stimulatory effect of TSH on the thyroid gland. The mass balance equations for plasma total T3 concentration, TT3P, and plasma total T4 concentration, TT4P, are:

$$\frac{d(TT3P)}{dt} = \frac{SRT3}{VP} + a_2 \cdot TT3F + a_3 \cdot TT3S - a_4 \cdot TT3P - a_7 \cdot TT3P$$

$$\frac{d(TT4P)}{dt} = \frac{SRT4}{VP} + a_{11} \cdot TT4F + a_{12} \cdot TT4S - a_{13} \cdot TT4P - a_{15} \cdot TT4P$$

(5.17)

where

TT3P = plasma total T3 concentration
TT3F = fast pool total T3 concentration
TT3S = slow pool total T3 concentration
TT4P = plasma total T4 concentration
TT4F = fast pool total T4 concentration
TT4S = slow pool total T4 concentration
VP = volume of plasma
a_2, a_3, etc., are constant parameters

An example using the thyroid model is described in Example 5.10.

Example 5.10

In this example, we used the thyroid hormone model described previously to simulate the effects of a temporary exposure to an endocrine-disrupting chemical (EDC) on the secretion rates of T3 and T4. To simplify the model, we did not write mass balance equations for the fast and slow pools, TT3F, TT3S, TT4F, and TT4S but assumed they were constant at their normal values. We used a time step of one hour and a time span of 0–30 hours. From hour 10 to hour 20, we reduced the parameters, k_1 and k_2, by a factor of 10, representing the EDC exposure. The function containing the differential equations and parameter values is as follows:

```
function Qdot = thyroid(t,Q)
%Constants for mass balance equations for TT3P and TT4P
a2 = 1.07; a3 = 0.292; a4 = 14.5; a7 = 0.0822;
a11 = 0.512; a12 = 0.501; a13 = 2.72; a15 = 0.1475;
% Normal value constants for fast and slow pools
TT3F = 10.65; TT3S = 2.34; TT4F = 231.0; TT4S = 223.0;
aHT = 4.1;
TRH = 2.98;                   %steady-state value of TRH
k1 = 0.269; k2 = 3.85; k3 = 131.0; k5 = 1.0;
kL = 0.77;                    %loss rate of TSH in plasma
K = 0.8;
VP = 3.05;                    %volumn of plasma
TSHnorm = 5.375;              %normal value of TSH
FT3Pnorm = 0.003*0.86;        %normal value of free plasma T3
                             %concentration
FT3P = .003*Q(2);             %free plasma T3 concentration

if t>=10 && t<=20
   k1 = 0.0269;
   k2 = 0.385;
end

%Rate of change of TSH, Q(1)
TSHdot = aHT*log(TRH)-k3*FT3P-k5*(FT3P-FT3Pnorm)-kL*Q(1);
% SRT3 = rate of secretion of T3
SRT3 = k1*Q(1)+K*(Q(1) - TSHnorm);
% SRT4 = rate of secretion of T4
SRT4 = k2*Q(1)+K*(Q(1) - TSHnorm);

%Rate of chamge of plasma T3, Q(2)
TT3Pdot = (SRT3/VP) + a2*TT3F + a3*TT3S - a4*Q(2) - a7*Q(2);
%Rate of change of plasma T4, Q(3)
TT4Pdot = a11*TT4F + a12*TT4S + (SRT4/VP) - a13*Q(3) - a15*Q(3);
Qdot = [TSHdot; TT3Pdot; TT4Pdot];
```

The m-file containing the d.e. solver and plot statements is as follows:

```
% Program m-file thyroid2 which calls function thyroid%
% Adapted from Saratchandran, et al. 1976.
% program to simulate effects of perchlorate on thyroid hormone system
%<----------------------------------------------------------->
% TT3P = plasma total T3 concentration, Q(2)
% TT3F = fast compartment total T3 concentration
% TT3S = slow compartment total T3 concentration
% TT4P = plasma total T4 concentration, Q(3)
% TT4F = fast compartment total T4 concentration
% TT4S = slow compartment total T4 concentration
% TSH = concentration of TSH in plasma
% SRT3 = rate of secretion of T3
% SRT4 = rate of secration of T4
% VP = volume of plasma
% a1, a2, etc., = various rate constants of material transfer
% kL = loss rate of TSH in plasma
% k1, k2, and K = parameters representing the stimulatory effect
%            of TSH on thyroid gland
% k3, k5 = parameters representing the inhibitory effect
%            of plasma free T3 on secretion rate of TSH
% aHT = parameter representing the stimulatory effect of
%            TRH on the rate of TSH secretion
%<----------------------------------------------------------->

tic

Q0 = [5.375; 0.86; 82.58];
tspan = [1:1:30];
[t,Q] = ode45('thyroid',tspan,Q0);

subplot(3,1,1)
plot(Q(:,1))
xlabel('Hours')
ylabel('THS Conc.')
subplot(3,1,2)
plot(Q(:,2))
xlabel('Hours')
ylabel('Plasma T3 Conc.')
subplot(3,1,3)
plot(Q(:,3))
ylabel('Plasma T4 Conc.')

toc
```

The results of the simulation are shown in Figure 5.31. The results confirm that reducing the secretion rates of T3 and T4, by making the parameters k_1 and k_2 smaller, leads to lower concentrations of plasma T3 and T4, thus increasing the concentration of TSH. Concentrations of plasma T3 and T4 reach new equilibria up to hour 20 when the values of k_1 and k_2 are returned to their original values. All concentrations then return to their original levels.

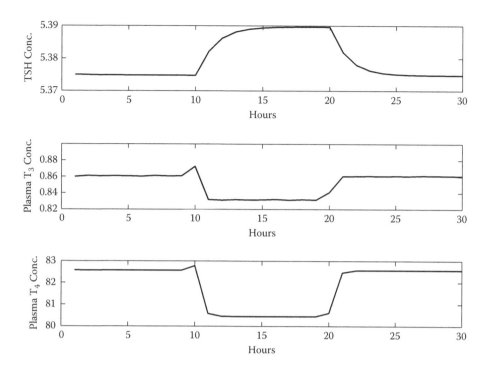

FIGURE 5.31   Plot of concentrations of TSH, plasma T3, and plasma T4 showing effects of endocrine dis-ruptor on secretion rate of T3 and T4.

EXERCISES

1. Modify the function **resp3** by making the exposure concentration, Y, a random vari-able with mean 3.0 and standard deviation 2.0. Use the normal random number generator, **randn.** Modify the m-file **resp4** to plot the resulting new values of the lung concentra-tion U and the exposure concentration Y. Use tic and toc to time the simulation. How does the simulation time compare with that for **resp3** and **resp4**?

2. In Example 5.2 we used script files **skin3** and **skin4** to simulate skin absorption with two parameters, the diffusion rate constant **k1** and the blood removal rate **k2**. Increase the diffusion rate constant from 0.05 to 0.10, 0.20, and 0.40. What do the results predict the effect is on external and percutaneous skin concentrations?

3. In Example 5.6, we simulated the transition between life stages exposed to a toxicant. In m-file **stagegrowth**, change the transition probabilities of remaining in life stages S1 and S2 from a linear response to a power function with base 10, starting at the normal value of 0.1 and ending at 1 at time step 5. Plot the new response and the resulting propor-tions in each life stage. Compare the stage proportions to those in Example 5.6.

REFERENCES

Albers, E. P., and K. R. Dixon. 2002. "A Conceptual Approach to Multiple-Model Integration in Whole Site Risk Assessment." In *Integrated Assessment and Decision Support. Proceedings of the First Biennial Meeting of the International Environmental Modelling and Software Society. Part 1,* ed. A. E. Rizzoli and A. J. Jakeman, 293–298. Manno, Switzerland: iEMSs.

Allen, L., and R. Ernest. 2002. "The Impact of Long-Range Dispersal on the Rate of Spread in Population and Epidemic Models." In *Mathematical Approaches for Emerging and Reemerging Infectious Diseases: An Introduction,* ed. C. Castillo-Chavez, S. Blower, P. van den Driessche, D. Kirschner, and A.-A. Yakubu, 183–197. New York: Springer.

ATSDR (Agency for Toxic Substances and Disease Registry). 2002. *Toxicological Profile for Aldrin/Dieldrin.* Public Health Service, Agency for Toxic Substances and Disease Registry, Atlanta, GA. http://www. atsdr.cdc.gov/toxprofiles/TP.asp?id=317&tid=56 (accessed 5/18/11).

Beck, J. R., and S. G. Paulker. 1983. "The Markov Process in Medical Prognosis." *Medical Decision Making* 3:419–458.

Carr, J. A., and C. Theodorakis. 2006. "Effects of Perchlorate in Amphibians." In *Perchlorate Ecotoxicology,* ed. R. J. Kendall and P. N. Smith, 127–153. Pensacola, FL: Society of Environmental Toxicology and Chemistry.

Carson, E., T. Hennessy, and A. Roudsari. 2001. "Control in Physiology and Medicine." In *Modelling Methodology for Physiology and Medicine,* ed. E. Carson and C. Cobelli, 15–44. New York: Academic Press.

Cooper, R. L., T. E. Stoker, J. M. Goldman, et al. 1996. "Effect of Atrazine on Ovarian Function in the Rat." *Reproductive Toxicology* 10:257–264.

Cooper, R. L., T. E. Stoker, J. M. Goldman, et al. 2000. "Atrazine Disrupts the Hypothalamic Control of Pituitary-Ovarian Function." *Toxicological Sciences* 53:297–307.

Dixon, K. R. 1977. "Thermal Plumes and Mercury Dynamics in Zooplankton." In *International Conference on Heavy Metals in the Environment, Vol. II, Part 2,* 875–886. Toronto, Ontario, Canada.

Dunn, J. E. 1978. "Optimal Sampling in Radio Telemetry Studies of Home Range." In *Time Series and Ecological Processes,* ed. H. H. Shugart, 53–70. Philadelphia, PA: SIAM Institute for Mathematics and Society.

Dunn, J. E., and P. S. Gipson. 1977. "Analysis of Radio Telemetry Data in Studies of Home Range." *Biometrics* 33:85–101.

Gottschalk, J. A. 1995. "Copper and Zinc Toxicity to the Gray Treefrog (*Hyla chrysocelis*) and the Northern Leopard Frog (*Rana pipiens*)." Master's thesis, Clemson University.

Hayes, W. J. Jr. 1974. "Distribution of Dieldrin following a Single Oral Dose." *Toxicology and Applied Pharmacology* 28:485–492.

Henriques, William. D. 1996. A Model of Spatial and Temporal Exposure and Effect of Dieldrin on Badgers at the Rocky Mountain Arsenal. Ph.D. dissertation, Clemson University.

Hoberg, J. R. 1993. *Atrazine Technical—Toxicity to the Freshwater Green Algae* (Selenastrum capricornutum). SLI Report # 93-4-4751. Wareham, MA: Springborn Laboratories, Inc.

Holgate, P. 1971. "Random Walk Models for Animal Behavior." In *Statistical Ecology, Vol. 2,* ed. G. P. Patil, E. C. Pielou, and W. C. Waters, 1–12. University Park: Pennsylvania State University Press.

Kendall, R. J., T. A. Anderson, R. J. Baker, et al. 2001. "Environmental Toxicology." In *Casarett and Doull's Toxicology: The Basic Science of Poisons. 6th Edition,* ed. C. D. Klaasen, 1013–1045. New York: McGraw-Hill.

Lefcort, H., R. A. Meguire, L. H. Wilson, et al. 1998. "Heavy Metals Alter the Survival, Growth, Metamorphosis, and Antipredatory Behavior of Columbia Spotted Frog (*Rana luteiventris*) Tadpoles." *Archives of Environmental Contamination and Toxicology,* 35:447–456.

Nichols, J. W., J. M., Anderson, M. E. Gargas, et al. 1990. "A Physiologically Based Toxicokinetics Model for the Uptake and Disposition of Waterborne Organic Chemicals in Fish." *Toxicology and Applied Pharmacology* 106:433–444.

Polishuk, Z. W., D. Wasserman, M. Wasserman, et al. 1977. "Organochlorine Compounds in Mother and Fetus in Labor." *Environmental Research* 13:278–284.

Reddy, M. B., R. S. H. Wang, H. J. Clewell III, and M. E. Andersen, eds. 2005. *Physiologically Based Pharmacokinetic Modeling.* Hoboken, NJ: Wiley.

Rees, S. E., S. Kjærgaard, and S. Andreassen. 2001. "Mathematical Modeling of Pulmonary Gas Exchange." In *Modelling Methodology for Physiology and Medicine,* ed. E. Carson and C. Cobelli, 253–278. San Diego, CA: Academic Press.

Saratchandran, P., E. R. Carson, and J. Reeve. 1976. "An Improved Mathematical Model of the Human Thyroid Hormone Regulation." *Clinical Endocrinology* 5:473–483.

Schaefer, H., and J. C. Jamoulle. 1988. "Skin Pharmacokinetics." *International Journal of Dermatology* 27:351–359.

Scheuplein, R. J., and I. H. Blank. 1971. "Permeability of the Skin." *Physiological Reviews*, 51:702–747.

Siniff, D. B., and C. R. Jessen. 1969. "A Simulation Model of Animal Movement Patterns." *Advanced Ecological Research* 6:185–219.

Solomon, K. R., J. P. Giesy, R. J. Kendall, et al. 2001. "Chlorpyrifos: Ecotoxicological Risk Assessment for Birds and Mammals in Corn Agroecosystems." *Human Ecological Risk Assessment* 7:497–632.

Sonnenberg, F. A., and J. R. Beck. 1993. "Markov Models in Medical Decision Making: A Practical Guide." *Medical Decision Making* 13:322–338.

Stacy, R. W. 1969. "The Comprehensive Patient-Monitoring Concept." In *Computers in Biomedical Research, Vol. III,* ed. R. W. Stacy and B. D. Waxman, 253–276. New York: Academic Press.

Tyler, J. A., and K. A. Rose. 1994. "Individual Variability and Spatial Heterogeneity in Fish Population Models." *Reviews in Fish Biology and Fisheries* 4:91–123.

Weiss, P., and J. L. Kavanau. 1957. "A Model of Growth and Growth Control in Mathematical Terms." *The Journal of General Physiology* 41:1–47.

Wolff, W. F. 1994. "An Individual-Oriented Model of a Wading Bird Nesting Colony." *Ecological Modelling* 72:75–114.

6 Modeling Ecotoxicology of Populations, Communities, and Ecosystems

Our focus to this point has been on an individual organism. It is essential to be able to extrapolate from effects of a contaminant on individuals to effects on populations. This extrapolation is complex for at least two reasons. First, the mortality caused by a pollutant to some individuals of a population may actually benefit survivors, because more resources such as food and habitat will be available to them. This phenomenon is known as *population-level compensation*. Second, many pollutant effects on individuals will be sublethal; that is, they will not lead directly to the death of the individual. Instead, they may decrease its efficiency in feeding, avoiding predators, and reproducing. Therefore, an assessment of the effects of a pollutant on a population will involve determining how various levels of individual impairment contribute to increased probabilities of mortality or reproductive failure through a variety of more direct factors.

6.1 EFFECTS OF TOXICANTS ON AGGREGATED POPULATIONS

Most aggregated population models ignore individual differences and use population parameters such as natality rates and survival rates. Aggregated models are best suited for simulation purposes other than prediction. They are particularly useful for developing hypotheses and evaluating management or policy alternatives.

One of the earliest population models is the logistic model of population growth published in 1838 by the doctor and mathematician, Pierre-François Verhulst (1838):

$$\frac{dN}{dt} = rN\left(1 - \frac{N}{K}\right) \tag{6.1}$$

where N is population size, r is the net population growth rate, and K is the carrying capacity or the maximum number of individuals that can be supported by the system resources. This is the same function used to model dose response and mortality (Equation 5, Table 2.1). The multiplier in parentheses reduces the rate of growth as N approaches carrying capacity. When N equals K, the rate of growth is zero. Without this multiplier, the population would grow exponentially. This continuous model applies to populations that grow continuously. Therefore, it is best used to model populations of small, usually invertebrate species. The dynamics of the modeled population are globally stable, with slight oscillations about the equilibrium K (Figure 6.1).

In the continuous model, the population increases rapidly until the carrying capacity is reached. The behavior of the discrete logistic model, however, is quite different (Figure 6.1) although both models use the same parameter values. This model,

$$N(t+1) = N(t) + rN(t)\left(1 - \frac{N(t)}{K}\right) \tag{6.2}$$

exhibits what has been called *chaotic behavior* (Li and Yorke 1975). In this type of dynamics, the behavior appears to be random but in fact is deterministic, with the degree of chaos determined by the initial values of the parameter r and initial population size N_0. These results illustrate the need for caution when using difference equations to approximate continuous processes that might better

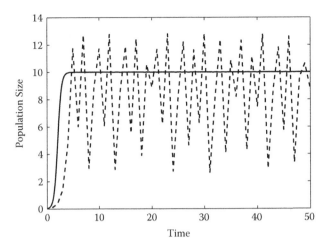

FIGURE 6.1   Plot of Verhulst continuous d.e. logistic growth model (solid line) and comparable difference equation (dashed line). Parameters in both models are identical.

be modeled using differential equations. The difference between differential and difference equations for the same process was explored by van der Vaart (1973).

Many population-level models do not track population numbers, but simulate the effects of acute exposure to a toxicant on an aggregated population. Before we can describe the dose response, however, we need to model mortality as a function of time. The relation between mortality, Y, and time, t, typically follows the logistic function:

$$Y = \frac{P_1}{1 + e^{(2.2/P_3)(P_2 - t)}}$$ (6.3)

where
   P_1 = maximum cumulative mortality
   P_2 = time at which 50% of individuals die
   P_3 = time between 10% and 50% mortality

Often, it is appropriate to use the mortality rate in a model, as opposed to the cumulative mortality. This is the derivative form of Equation 5 in Table 2.1 (see Chapter 2). In terms of Equation (6.3), the derivative is:

$$\frac{dY}{dt} = \frac{2.2}{P_1 P_3} Y (P_1 - Y)$$ (6.4)

Note that the parameter, P_2, and the independent variable, time, do not appear explicitly in the equation. This allows the mortality rate equation to be used in differential equations without tying mortality to a specific time relative to the start of the simulation.

Example 6.1

In this example, we compare the cumulative mortality function and the numerical solution to Equation (6.4) using data on mortality in *Daphnia magna* following exposure to chlorpyrifos from Naddy et al. (2000). The parameters for the model are obtained using nonlinear regression, as shown in Example 7.4. The MATLAB function **daphniapop** contains the d.e. and the m-file **daphniapop2** contains the d.e. solver and the plot statements:

```
% function daphniapop
% m-file function defining parameters and d.e. for
% logistic function of mortality
function mdot = daphniapop(~,m)
p1=20.0;
%p2=11.7195;
p3=2.3147;
b = 2.2./(p3*p1);
mdot = b*m.*(p1-m);


% m-file daphniapop2
% program to solve logistic d.e. of mortality in m-file daphniapop

day=[0.0420 0.1250 0.25 0.50 1.00 1.50 2.00 2.50 3.00 3.50...
     4.00 5.00 6.00 7.00 8.00 9.00 10.00 11.00 12.00 13.00...
     14.00 15.00 16.00 17.00 18.00 19.00 20.00 21.00];
mort=[0 0 0 0 0 0 0 0 0 0 0 0 0 0 0 0 1 6 6 10 15 20 20 20 20 20 20
20]';
p1=20.0;
%p2=11.7195;
p3=2.3147;
y=p1./(1 + exp((2.2/p3)*(p2-t)));
m0=0.0003;
tspan = [0 25];
[t,m] = ode45('daphniapop', tspan, m0);

plot(t,m,'k-','LineWidth',2)
hold on
xlabel('Time (days)')
ylabel('Daphnia Population')

plot(t,y,'k--','LineWidth',2)
legend('solution to d.e.','logistic plot')
plot(day,mort,'o','MarkerSize',6)
```

The results of the MATLAB routines are shown in Figure 6.2. Note that the curves are virtually identical. Although the solution to the differential equation does not depend on time, it does depend on the initial value of mortality. We adjusted the initial value to obtain the fit to the data by eye.

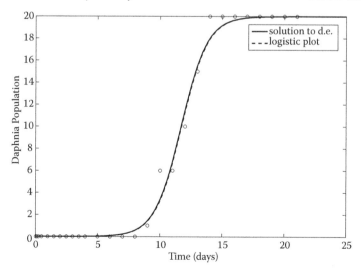

FIGURE 6.2   Plot of solution to d.e. (solid line) and logistic mortality function (dashed line). (Data from R. Naddy et al., 2000. "Response of *Daphnia magna* to Pulsed Exposures of Chlorpyrifos." *Environmental Toxicology and Chemistry* 19, 2, 423–431. With permission from John Wiley and Sons.)

The example above shows the use of the logistic function to model mortality, but only for a single independent variable (concentration). Mortality also can be a function of more than one independent variable. We need to fit the logistic to data for a range of values for each independent variable to predict the level of mortality as a function of those variables. These data also will provide the basis for the dose–response relationship. The steps involved in developing this model are:

1. Fit a (logistic) model to the mortality data as a function of time for each exposure concentration using least squares regression (see Chapter 7).
2. Choose a model(s), based on a plot of the data, to describe the relation between the model parameters and the concentrations.
3. Fit the model to the parameter/concentration data using least squares regression.
4. Substitute the equations obtained in Step 3 for the parameters in the (logistic) model.

The model is now capable of predicting mortality for the several variables.

Example 6.2

In this example, striped bass egg mortality was modeled to simulate exposure to electric power condenser entrainment and effluent discharge conditions (Hall et al. 1981). The independent variables influencing mortality were total residual chlorine (TRC), elevated temperatures (ΔT), and exposure time (EXP). Figure 6.3 shows a plot of the mean and 95% confidence intervals along with the predicted curve of the final model. The simulated conditions in this plot are TRC = 0.15mg/l, ΔT = 6°C, and EXP = 2.0 h. The model was implemented following the steps described above. First, the logistic model was fit to the data of egg mortality over time for each combination of TRC, ΔT, and EXP using nonlinear least squares regression to estimate the parameters P_1, P_2, and P_3 (Table 6.1).

Second, the parameter values derived for each combination of independent variables are regressed against the respective independent values:

$$P_1 = -6.73 + 176.25\,TRC + 2.42\,\Delta T + 8.71EXP$$

$$P_2 = 18.66 - 0.583\,TRC - 0.260\,\Delta T + 0.278EXP \qquad (6.5)$$

$$P_3 = 28.66 - 22.26\,TRC - 0.668\,\Delta T - 2.805EXP$$

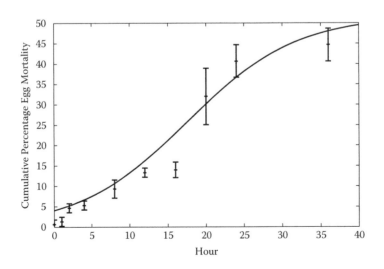

FIGURE 6.3 Observed (error bars) and predicted (smooth curve) striped bass egg mortality following exposure to simulated power plant conditions. TRC = 0.15mg/l, ΔT = 6°C and 2.0 h exposure. (Data from L. Hall, et al., 1981. "Time-Related Mortality Responses of Striped Bass (*Morone saxatilis*) Ichthyoplankton after Exposure to Simulated Power Plant Chlorination Conditions." *Water Research* 15, 903–910. With permission from Elsevier.)

TABLE 6.1
P_1, P_2, and P_3 Values at Various Exposure Conditions for Striped Bass Eggs from Fitting the Logistic Model to Data on Mortality over Time

Exposure Time (h)	TRC (mg/l)	ΔT (°C)	P_1	P_2	P_3	r^2
0.08	0.00	6	22.832	17.526	27.700	0.956
	0.15	6	27.374	13.681	16.478	0.977
	0.15	10	32.970	15.895	19.507	0.971
	0.30	2	42.020	19.441	23.938	0.975
2.0	0.00	2	13.916	19.749	26.130	0.905
	0.00	6	32.528	15.792	15.829	0.982
	0.15	2	51.802	20.580	20.351	0.930
	0.15	6	47.644	17.327	11.337	0.940
	0.15	10	64.648	18.727	9.606	0.971
	0.30	6	86.952	17.258	11.266	0.966
	0.30	10	92.298	16.796	10.915	0.959
4.0	0.00	10	48.909	17.640	14.291	0.942
	0.15	2	59.406	18.942	7.880	0.956
	0.15	6	57.848	17.970	9.870	0.928
	0.30	6	100.000	17.311	12.913	0.941
	0.15	0	42.326	19.589	13.494	0.948
	0.30	0	89.382	18.803	8.960	0.937

Source: Data from L. Hall, et al., 1981. "Time-Related Mortality Responses of Striped Bass (*Morone saxatilis*) Ichthyoplankton after Exposure to Simulated Power Plant Chlorination Conditions." *Water Research,* 15:903–910.

And finally, the equations for parameters P_1, P_2, and P_3, are substituted into the logistic model, which now can be used to predict egg mortality for any combination of TRC, ΔT, and EXP within the range measured in the mortality experiments (Figure 6.3). To iterate, the output in the figure is not a fit of the model to the data. It is the solution to the differential equation (Equation 6.4) with Equation (6.5) substituted for the model parameters.

6.2   EFFECTS OF TOXICANTS ON AGE-STRUCTURED POPULATIONS

Age-structured models have a long and significant history in population modeling. The most common and widely applied model is the age-distribution vector model (Bernardelli 1941, Lewis 1942, Leslie 1945, 1948). The matrix approach was extended to life stages by Lefkovitch (1965). This modeling approach is similar to the transition matrix approach described in Section 5.1.4.2 with the addition of a reproductive vector. This matrix, however, does not meet the criteria for a Markov process.

The age-distribution vector, usually limited to females, can be described by the vector, n_t:

$$n_t = [v_{0t} \quad v_{1t} \quad v_{1t} \quad \cdots \quad v_{it} \quad \cdots \quad v_{kt}] \qquad (6.6)$$

where v_{it} is the number of individuals aged i at time t. We would like to obtain a new age distribution vector at time $t + 1$, v_{it+1}. This new vector is obtained by premultiplying the vector v_{it} by the matrix M:

$$n_{t+1} = \begin{bmatrix} v_{0,t+1} \\ v_{1,t+1} \\ v_{2,t+1} \\ \vdots \\ v_{k,t+1} \end{bmatrix} = \begin{bmatrix} f_0 & f_1 & f_2 & \cdots & f_k \\ s_0 & 0 & 0 & \cdots & 0 \\ 0 & s_1 & 0 & \cdots & 0 \\ \vdots & & \ddots & \ddots & \vdots \\ 0 & 0 & 0 & s_{k-1} & 0 \end{bmatrix} \begin{bmatrix} v_{0,t} \\ v_{1,t} \\ v_{2,t} \\ \vdots \\ v_{k,t} \end{bmatrix} = Mn_t \tag{6.7}$$

where

f_i = the average number of female offspring born per female between time t and $t + 1$ that survive to time $t + 1$, and

s_i = the probability that a female aged i at time t survives to time $t + 1$

The main difference between this matrix model and a Markov model is that in the Markov model, $f_i + s_i = 1$ for each v_{it}, whereas in the matrix model, these sums can be greater or less than 1. As in the Markov model, the proportion in each age class may converge on a stable value, the stable age distribution, after some number of iterations if the parameters f_i and s_i remain constant. The other possible outcome would be oscillating proportions. Also like the Markov model, the matrix model can be expressed by a system of equations by expanding the matrix multiplication.

The total number of female offspring born into the $v_{0,t+1}$ age class is:

$$v_{0,t+1} = f_0 v_{0,t} + f_1 v_{1,t} + f_2 v_{2,t} + \cdots + f_k v_{k,t} = \sum_{i=0}^{k} f_i v_{i,t} \tag{6.8}$$

The older age classes are:

$$v_{1,t+1} = s_0 v_{0,t}$$
$$v_{2,t+1} = s_1 v_{1,t}$$
$$\vdots \tag{6.9}$$
$$v_{k,t+1} = s_{k-1} v_{k-1,t}$$

Because the parameters f_i and s_i are not constant for our purposes, but are functions of toxic substances, the system of equations may be easier to apply than the matrix structure.

Example 6.3

An example of a population projection matrix model is the stage-based model of amphibian populations developed by Vonesh and De la Cruz (2002), which was used in an environmental toxicology context by Cox et al. (2006). We modified the model to simulate five life stages: eggs, larvae, metamorphs, juveniles, and adults. Although there are many ways toxicants can affect amphibian reproduction and survival, in this example we examine the effects of an endocrine-disrupting chemical (EDC) on the probability of metamorphosis occurring in the larval stage. Survival and reproduction in real-world populations will be functions of environmental variables; however, we used the constant parameters of Vonesh and De la Cruz (2002). The equations for this model are:

$$E_{t+1} = \rho \cdot \phi \cdot A_t$$

$$L_{t+1} = \sigma_e (1 - P\text{met}) E_t$$

$$M_{t+1} = \sigma_l \cdot P\text{met} \cdot L_t \qquad (6.10)$$

$$J_{t+1} = \sigma_m (1 - P) M_t$$

$$A_{t+1} = \sigma_j \cdot P \cdot J_t$$

where E, L, M, J, and A refer to eggs, larvae, metamorphs, juveniles, and adults, respectively, and

ρ = adult sex ratio
ϕ = per capita egg production
σ_e = egg survival rate
σ_l = larval survival rate
σ_m = metamorph survival rate
σ_j = juvenile survival rate
Pmet = larval rate of metamorphosis
P = larval maturation probability

The larval survival rate, σ_l, is a function of larval density:

$$\sigma_l = \frac{\sigma_{lmax}}{(1 + d \cdot L_t)^\gamma} \qquad (6.11)$$

where

σ_{lmax} = maximal larval survival in the absence of density dependence
d = density-dependence coefficient
γ = density-dependence exponent

Because the parameters are constant, we know that likely there will be a stable-stage distribution. Following initial transient dynamics, this distribution was nearly reached after about 10 years (Figure 6.4). To simulate the effect of an EDC on metamorphosis, we reduced the probability of metamorphosis from 0.9 to 0.5 after 15 years. The results show an increase in the larval density, as fewer larvae metamorphose, and a concomitant decrease in the density of metamorphs, larvae, and adults. The stage distribution again stabilizes at new densities (Table 6.2). The greatest change in stage-class size is in the larvae stage, reflecting the decrease in the rate of larval metamorphosis.

We used this example to illustrate the basic structure of an age (stage)-distribution vector model. Obviously, a realistic predictive model would have all parameters as functions of environmental variables, which exhibit inherent variability.

6.3 EFFECTS OF TOXICANTS ON COMMUNITIES

In Section 6.2 we looked at aggregated models of populations. That is, models of an aggregation of individuals of the same species. In this section, we examine models of communities. A community can be defined as an assemblage of different species.

Modeling communities implies that the model simulates some relationship between species, such as with predators and their prey. Multiple species in a system constitute a food web. As with population models, most food web models are aggregated, as opposed to individual based. Individual-based food web models necessarily would be highly complex with a large number of parameters. Food web models may have a single trophic level (e.g., herbivores) or more than one trophic level (e.g., plants, herbivores, carnivores).

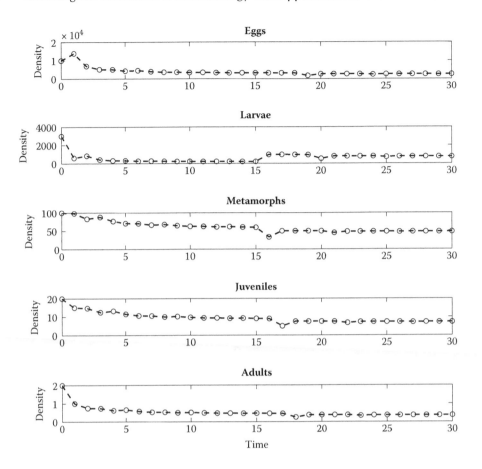

FIGURE 6.4   Simulated densities of *Bufo boreas* life stages before and after exposure to hypothetical endocrine disrupting chemical.

TABLE 6.2
Stable Stage Distributions with and without Exposure to an EDC

	Life Stage				
	Eggs	**Larvae**	**Metamorphs**	**Juveniles**	**Adults**
Nonexposed	2989	179	57.2	8.59	0.43
Exposed	2506	752	48.0	7.20	0.36

Some of the earliest multispecies population models were aggregated predator-prey models (Lotka 1923, Volterra 1926). A slightly more complex model, which illustrates an aggregated predator-prey model, is described below (Dixon and Cornwell 1970).

$$N(t+1) = N(t) + b \cdot N(t) - N_a(t)$$

$$P(t+1) = P(t) + P_b(t) - d \cdot P(t)$$

(6.12)

where

$N(t+1)$ = prey population at time $t + 1$
$N(t)$ = prey population at time t
b = prey birth rate
$Na(t)$ = number of prey attacked during time t to $t + 1$
$P(t+1)$ = predator population at time $t + 1$
$P(t)$ = predator population at time t
$Pb(t)$ = number of prey born during time t to $t + 1$
d = predator death rate

The number of prey attacked during time t to $t + 1$, Na, is expressed by the equation:

$$Na(t) = P(t) \cdot K_1 \left(1 - e^{-aN(t)^2 P(t)^{(1-c)}} \right)$$

(6.13)

where

K_1 = maximum number of attacks per predator
a and c = constants.

The equation for the number of predators born during time t to $t + 1$, Pb, has the same structure as that for Na:

$$Pb(t) = P(t) \cdot K_2 \left(1 - e^{-fN(t)^2 P(t)^{(1-g)}} \right)$$

(6.14)

where

K_2 = maximum number of offspring produced per predator
f and g = constants.

Example 6.4

This example is based on the above model, which was used to simulate the wolf and moose populations on Isle Royale, Michigan (Dixon and Cornwell 1970). The two populations are in equilibrium at 571 moose and 28 wolves after some initial transient dynamics (Figure 6.5). Because the two populations are relatively small, we rounded the population numbers to the next smallest integer value using the MATLAB function **floor**.

At equilibrium, the number of moose attacked is balanced by the number of moose born. Because the moose population does not greatly fluctuate, the wolf population also stays at equilibrium. The problem in using a simple aggregated model is that all individuals are assumed to be identical. There is no age or sex assigned to the individuals in the population. Because the model uses a time step of a year, and no interannual dynamics are modeled, we used a symbol for each time step connected by a dashed line rather than a continuous line to display the output.

Despite its limitations, the model can illustrate the effect of changing the parameters in the model. Let's assume that a contaminant has increased the wolf death rate d by 50%. These results are shown in Figure 6.6. In this simulation, the increased wolf death rate decreases the wolf population, allowing the moose population to increase. With an increased food supply, the number of wolves born increases and the wolf population recovers. Both moose and wolf numbers show classic damped oscillations until they approach new equilibria.

The MATLAB m-file for this model is:

```
% m-file wolfmoose
% Difference equation model of predator and prey
N = 571.;P = 28;K1 = 45.45;K2 = 1.6;
```

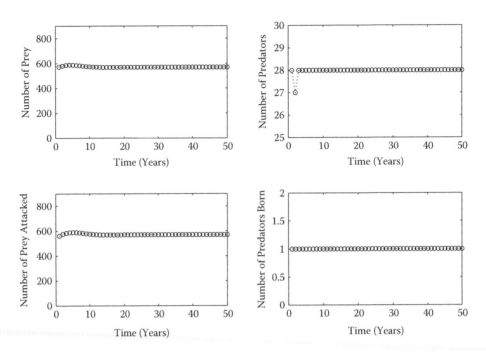

FIGURE 6.5 Output from wolf-moose model showing number of moose, wolves, moose attacked, and wolves born, using normal parameters.

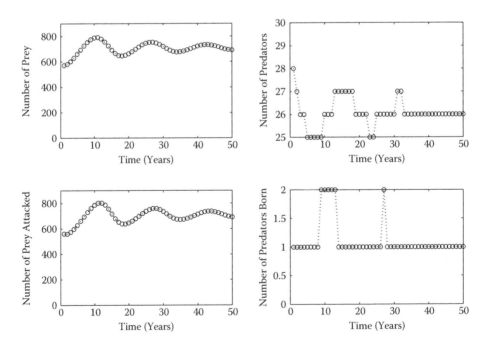

FIGURE 6.6 Output from wolf-moose model showing numbers of moose, number of wolves, number of moose attacked, and number of wolves born, with wolf mortality increased 50%.

```
a = 7.760e-7;f = 2.7136e-8;
c = 0.75;g = 0.65;
b = 1.0;
d = 0.045;
%d = 0.0675;

xmax = 50;
x = 1:1:xmax+1;
z = 1:1:xmax;
for t = 1:xmax
    Na(t) = P(t)*K1*(1-exp(-a*(N(t).^2)*(P(t).^(1-c))));
    Nar = floor(Na);
    Pb(t) = P(t)*K2*(1-exp(-f*(N(t).^2)*(P(t).^(1-g))));
    Pbr = floor(Pb);
    N(t+1) = N(t) + b*N(t) - Na(t);
    if N(t+1) < 0
        N(t+1) = 0;
    end
    Nr = floor(N);
    P(t+1) = P(t) + Pb(t) - d*P(t);
    Pr = floor(P);
end

hold on
subplot(2,2,1), plot(x,Nr,':o')
axis([0,xmax,0,900])
xlabel('Time (years)','fontsize',12)
ylabel('Number of Prey','fontsize',12)
subplot(2,2,2), plot(x,Pr,':o')
axis([0,xmax,25,30])
xlabel('Time (years)','fontsize',12)
ylabel('Number of Predators','fontsize',12)
subplot(2,2,3), plot(z,Nar,':o')
axis([0,xmax,0,900])
xlabel('Time (years)','fontsize',12)
ylabel('Number of Prey Attacked','fontsize',12)
subplot(2,2,4), plot(z,Pbr,':o')
axis([0,xmax,0,2])
xlabel('Time (years)','fontsize',12)
ylabel('Number of Predators Born','fontsize',12)
```

6.4 EFFECTS OF TOXICANTS ON ECOSYSTEMS

In this section, we examine models of ecosystems. An *ecosystem* is defined as a community inter-acting with nonliving components such as nutrients, temperature, and precipitation.

The cycling of toxic elements and compounds determines the long-term concentrations of the elements or compounds in the environment, which will determine the exposure of plants and animals. In addition, environmental contamination with toxic substances may interfere with ecosystem processes such as nutrient cycling and litter decomposition. For example, in a study of the effects of heavy metals on litter decomposition on the Crooked Creek watershed in Missouri, Watson et al. (1976) found that concentrations of lead, zinc, copper, and cadmium were posi-tively correlated with litter biomass and nutrients, suggesting that the metals had lowered the rate of decomposition.

To investigate cycling of cesium on a forest floor ecosystem, Patten and Witkamp (1967) used laboratory microcosms with litter, soil, microflora, millipedes, and leachate compartments with the litter spiked with radioactive cesium, [134]Cs (Figure 6.7).

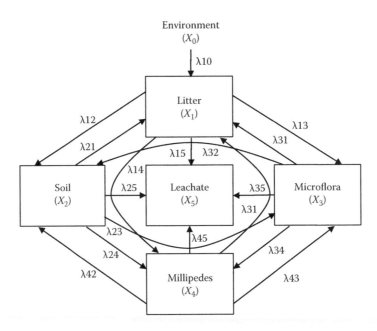

FIGURE 6.7 Flow diagram of litter-soil-microflora-millipede-leachate microecosystem showing all possible routes of cesium transfer and associated rate constants. (From B. C. Patten and M. Witkamp. 1967. "Systems Analysis of ^{134}Cesium Kinetics." *Ecology* 48, 5, 813–824. With permission from Ecological Society of America.)

The model assumed first-order kinetics. That is, the transfer from compartment i to compartment j is proportional to the amount of cesium in compartment i, and the transfer coefficients, λ_{ij}, are constant with units t^{-1}. The rate of change in the ^{134}Cs in the j_{th} compartment will be the difference in the sum of the sources of input into the compartment and the sum of the losses out of the compartment. Although some of the transfer coefficients in the microecosystem were zero, the model can be described by the system of ordinary, linear, constant coefficient differential equations:

$$\frac{dx_j}{dt} = \sum_i \lambda_{ij}x_i - \sum_j \lambda_{ji}x_j \tag{6.15}$$

where x_j is the amount of ^{134}Cs in compartment j. The following example solves these equations using both MATLAB and Simulink.

Example 6.5

This example is taken from the paper by Patten and Witkamp (1967) using their microcosm combination V with compartments 1 (litter), 3 (microflora), 4 (millipedes), and 5 (leachate) (see Figure 6.7). The m-files to solve the system of Equations (6.15) are cesiumV and cesiumsolver. CesiumV contains model parameters and differential equations. Cesiumsolver contains the initial conditions, simulation time, the d.e. solver, and plot statements.

```
%function cesiumV
%Combination V of Patten and Witkamp (1967)
%Compartments are litter (x1), soil (x2), microflora (x3),
%millipedes (x4), and leachate (x5)

function xdot=cesiumV(~,x)
%rate constants
l01=0.0;
l12=0.0; l13=0.065; l14=0.020; l15=0.025;
l21=0.0; l23=0.0; l24=0.0; l25=0.0;
l31=0.0; l32=0.0; l34=0.375; l35=0.0;
l41=0.475; l42=0.0; l43=0.050; l45=0.0;

%d.e. for litter compartment, x1
x1dot=l01+(-(l12+l13+l14+l15)*x(1))+l21*x(2)+l31*x(3)+l41*x(4);
%d.e. for soil compartment, x2
x2dot=(l12*x(1))-(l21+l23+l24+l25)*x(2)+l32*x(3)+l42*x(4);
%d.e. for microflora compartment, x3
x3dot=l13*x(1)+l23*x(2)-(l31+l32+l34+l35)*x(3)+l43*x(4);
%d.e. for millipede compartment, x4
x4dot=l14*x(1)+l24*x(2)+l34*x(3)-(l41+l42+l43+l45)*x(4);
%d.e. for leachate compartment, x5
x5dot=l15*x(1)+l25*x(2)+l35*x(3)+l45*x(4);
%vector of differential equations
xdot=[x1dot;x2dot;x3dot;x4dot;x5dot];

%m-file cesiumsolver
x0=[1.0 0.00 0.00 0.00 0.00]';        %vector of initial conditions
tspan=(0:.1:30);
[t,x]=ode45(@cesiumV,tspan,x0);       %differential equation solver

plot(t,x(:,1),'r');
hold on;
plot(t,x(:,3),'k');
plot(t,x(:,4),'m');
plot(t,x(:,5),'b');
xlabel('Time(days)');
ylabel('y_j = x_j/X');
title('Radiocesium in microcosms')
legend1=legend({'litter','microflora','millipedes',...
        'leachate'},'Position',[0.5901 0.5647 0.1607 0.1889]);
```

The results (Figure 6.8) show a decline in the cesium concentration in the litter compartment (the compartment in which the cesium was introduced). As the cesium in the microflora and millipede compartments increase and then level off, the rate of loss from the litter compartment deceases. The cesium in the leachate compartment continues to increase as there is no loss (negative) term in the differential equation for the leachate compartment.

The block diagram for the cesiumV model is shown in Figure 6.9. All four of the state variables (cesium in the four compartments) are input into a mux (multiplexer) block to display them in the scope and copied into the Workspace. Double-click the **Simout** block and select **Array** in the **Save format** line. To run the simulation, choose **Simulation > Start** on the model toolbar. Typing the command **plot(Simout)** plots the results of the output signal that was input into the **To Workspace** block (Figure 6.10). The plot is edited using the **Edit Plot** button in the figure window to add labels and increase font size and line width. Notice the similarity between Figures 6.8 and 6.10.

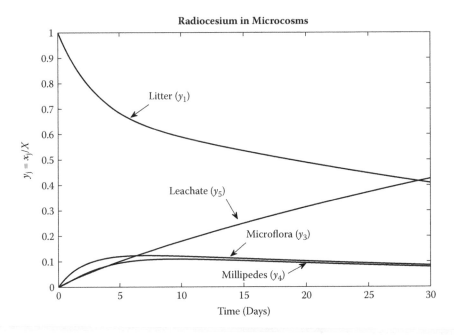

FIGURE 6.8 Radiocesium in microcosms with litter, microflora, millipede, and leachate compartments. (Adapted from B. C. Patten and M. Witkamp. 1967. "Systems Analysis of ^{134}Cesium Kinetics." *Ecology* 48, 5, 813–824. With permission from Ecological Society of America.)

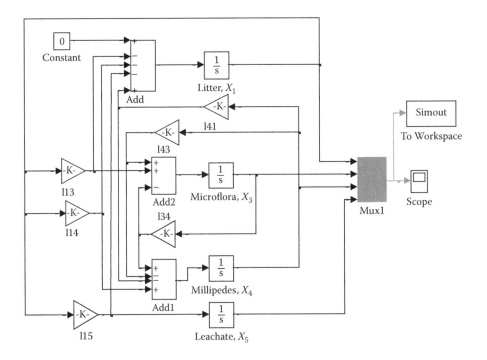

FIGURE 6.9 Block diagram for the ^{134}Cesium microcosm model.

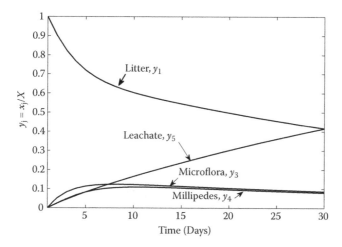

FIGURE 6.10 Output from Simulink model Cesium5 showing radiocesium in microcosm with litter, microflora, millipede, and leachate compartments. (Adapted from B. C. Patten and M. Witkamp. 1967. "Systems Analysis of ^{134}Cesium Kinetics." *Ecology* 48, 5, 813–824. With permission from Ecological Society of America.)

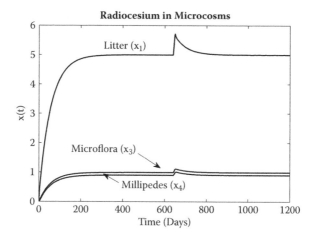

FIGURE 6.11 Simulated cesium concentration with constant input and perturbation after equilibrium. (Adapted from B. C. Patten and M. Witkamp. 1967. "Systems Analysis of ^{134}Cesium Kinetics." *Ecology* 48, 5, 813–824. With permission from Ecological Society of America.)

EXERCISES

1. Open m-file **cesiumsolver** and function **cesiumV**.
2. Modify **cesiumsolver** to plot only state variables litter, microflora, and millipedes as these variables are the only ones reaching equilibrium. Save the file under a different name.
3. Modify function **cesiumV** to increase the cesium input λ_{01} from 0.125 to 0.250 between days 640 and 648. Refer to Patten and Witkamp (1967).
4. Run the new **cesiumsolver** to show the results of the disturbance in cesium input. Compare your results with Figure 6.11 (see also Patten and Witkamp 1967, Figure 4). Labels were added to the compartment output plots using Edit Plot and the Property Editor.
5. Increase λ_{01} further and observe the results.

REFERENCES

Bernardelli, H. 1941. "Population waves." *Journal of the Burma Research Society* 31:1–18.

Cox, S. B., K. R. Dixon, and R. J. Kendall. 2006. "Application of Population Models to Perchlorate Ecotoxicology." In *Perchlorate Ecotoxicology*, ed. R. J. Kendall and P. N. Smith, 211–227. Pensacola, FL: Society of Environmental Toxicology and Chemistry.

Dixon, K. R., and G. W. Cornwell. 1970. "A Mathematical Model for Predator and Prey Populations." *Researches on Population Ecology* 12:127–136.

Hall, L. W., Jr., D. T. Burton, S. L. Margrey, and K. R. Dixon. 1981. "Time-Related Mortality Responses of Striped Bass (*Morone saxatilis*) Ichthyoplankton after Exposure to Simulated Power Plant Chlorination Conditions." *Water Research* 15:903–910.

Lefkovitch, L. P. 1965. "The Study of Population Growth in Organisms Grouped by Stages." *Biometrics* 21:1–18.

Leslie, P. H. 1945. "The Use of Matrices in Certain Population Mathematics." *Biometrika* 33:183–212.

Leslie, P. H. 1948. "Some Further Notes on the Use of Matrices in Population Mathematics." *Biometrika* 35:213–245.

Lewis, E. G. 1942. "On the Generation and Growth of a Population." *Sankhya* 6:93–96.

Li, T. Y., and J. A. Yorke, 1975. "Period Three Implies Chaos." *American Mathematical Monthly* 82:985–992.

Lotka, A. J. 1923. "Contribution to Quantitative Parasitology." *Journal of the Washington Academy of Sciences* 13:152–158.

Naddy, R. B., K. A. Johnson, and S. J. Klaine. 2000. "Response of *Daphnia magna* to Pulsed Exposures of Chlorpyrifos." *Environmental Toxicology and Chemistry* 19:423–431.

Patten, B. C., and M. Witkamp. 1967. "Systems Analysis of ^{134}Cesium Kinetics." *Ecology* 48:813–824.

van der Vaart, H. R. 1973. "A Comparative Investigation of Certain Difference Equations and Related Differential Equations: Implications for Model Building." *Bulletin of Mathematical Biology* 35:195–211.

Verhulst, P.-F. 1838. "Notice sur la loi que la population suit dans son accroissement." *Correspondance Mathématique et Physique* 10:113–121.

Volterra, V. 1926. "Fluctuations in the Abundance of a Species Considered Mathematically." *Nature* 118:558–560.

Vonesh, J. R., and O. De la Cruz. 2002. "Complex Life Cycles and Density Dependence: Assessing the Contribution of Egg Mortality to Amphibian Declines." *Oecologia* 133:325–333.

Watson, A. P., R. I. Van Hook, D. R. Jackson, and D. E. Reichle. 1976. *Impact of a Lead Mining-Smelting Complex on the Forest-Floor Litter Arthropod Fauna in the New Lead Belt Region of Southeast Missouri.* ORNL/NSF/EATC-30. Oak Ridge, TN: Oak Ridge National Laboratory.

7 Parameter Estimation

Parameter estimation can be considered the process of obtaining numerical values for parameters in model equations. Model parameters need "good" values. Otherwise, model predictions will have great uncertainty. Good values come from good data. As pointed out in Section 1.4.1, the adage, "Garbage In, Garbage Out," has a lot of truth. Statistical methods of estimation provide the "best" parameter values.

There are many statistical methods of estimating parameters. Two of the more common types used in parameter estimation are *linear least squares regression* and *nonlinear least squares regression*. Both linear and nonlinear regression can be used to estimate parameters in functions in which the relationship between the dependent and independent variables is nonlinear. There are advantages and disadvantages to each method and each is more appropriate for different types of models. We will not go into the statistical theory of regression here. The theory of regression is covered in many standard texts, some of which are listed at the end of this chapter.

Whether you use linear or nonlinear regression, there are several steps in the process:

1. Plot the data to visually determine the relationship between the independent and dependent variables.
2. Based on the relationship, select the model to fit to the data.
3. Select linear or nonlinear regression.
4. Fit the model to the data using regression.
5. Determine if the fit is adequate.

7.1 LINEAR REGRESSION

Linear least squares regression is the most widely used statistical method of regression. This is a result largely of the long history of this method, having been developed in the late eighteenth century by Karl Friedrich Gauss. Given an adequate data set, linear least squares regression can be used to fit a model of the form:

$$y = \beta_0 + \beta_1 x_1 + \beta_2 x_2 + ... \tag{7.1}$$

to the data, where y is the response variable, the x_i are the explanatory variables, and the β_i are the parameters. The regression is linear because each term in the equation is linear in the parameters, not the explanatory variables. For example, the following equation also is linear in the parameters, or is *statistically linear*:

$$y = \beta_0 + \beta_1 x_1^2 + \beta_2 \ln(x_2) + \beta_3 \sin x_3 \tag{7.2}$$

It should be clear that an equation that is linear in the parameters can model a nonlinear function, not just a straight line. Simple linear regression (with one explanatory variable and one response variable), however, is best used to fit a straight-line relationship.

Nonlinear regression is used when the relationship is nonlinear in the parameters such as β^2, $\log \beta$, and $e^{\beta x}$. Often, nonlinear models have been linearized using various transformations to avoid using nonlinear regression. This is a result of not having computers available in the late eighteenth

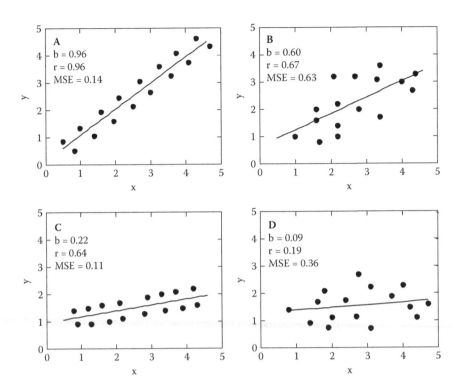

FIGURE 7.1   Examples of regression and their goodness-of-fit statistics. (Adapted from K. E. F. Watt. *Ecology and Resource Management: A Quantitative Approach*, New York: McGraw-Hill, 1968. With permission).

century and historical inertia. With today's computing power and statistical software, this no longer is a valid reason not to use nonlinear regression when fitting a nonlinear model. Depending on the model, linearization can introduce significant bias in the parameter estimates. Several diagnostic statistics describe how well the model fits the data. Three of the most important are the slope of the regression line, b, the correlation coefficient, r, and the mean squared error, MSE. Although the slope and correlation are most often reported, the MSE is just as important in evaluating the fit of the regression and should be reported along with b and r. The reason can be seen in Figure 7.1. The MSE, which measures the scatter about the regression line, is good (i.e., low scatter) in A and C in the figure. It is relatively high in B and D. The slope, of course, measures the strength of the relationship between x and y. It is greater in A and B than in C and D. The correlation coefficient actually is less informative because it is a function of both b and MSE. Therefore, the picture is less clear when looking only at r. In A, it is high because b is also high and MSE is low. In D, r is low because b is low and MSE is somewhat higher than in A and C. In B and C, r is about equal because, even though b is greater in B than C, this is offset by MSE being lower in C than in B.

Three MATLAB linear regression functions are described below: **regress**, **polyfit**, and **regstats**.

7.1.1   FUNCTION: REGRESS

Multiple linear regression provides a least squares fit of models such as Equation (7.1) above, using the **regress** MATLAB function:

```
[b,bint,r,rint,stats] = regress(y,X,alpha)
```

by solving the linear model

$$y = X\beta + \varepsilon \tag{7.3}$$

for β where y is an n-by-1 vector of responses, X is an n-by-p matrix of explanatory variables, β is a p-by-1 vector of parameters, and ε is an n-by-1 vector of random errors that are assumed to be normally distributed with equal variances. The estimates of β are returned in **b** and $100(1 - \text{alpha})\%$ confidence intervals for the estimates in **bint**. For example, alpha = 0.05 gives 95% confidence intervals. Residuals are returned in the n-by-1 vector **r** and $100(1 - \text{alpha})\%$ confidence intervals in the n-by-2 vector **rint**. The vector **stats** contains the R^2 statistic along with the F and p values for the regression, and an estimate of the error variance. By limiting the matrix X to a single explanatory variable, regress will fit a straight line to the data. Note that X must contain a column of ones to estimate the intercept, β_0.

Example 7.1

This example fits a simple linear regression model to LD_{50} data as a function of body weight in birds. The data in Table 7.1 are selected from Solomon et al. (2001, Table 34). To obtain the results from **regress**, the m-file **regressLD50vswt** was used:

```
%m-file regessLD50vswt
%program to do linear regression with MATLAB function regress

X = [1 28
     1 52
     1 82
```

TABLE 7.1
LD_{50} Values by Body Weight for 11 Species of Birds

Species	Body Weight (g)	LD_{50} (mg/bird)
House sparrow	28	0.83
Red-winged blackbird	52	0.69
European starling	82	6.15
Japanese quail	90	1.40
Common grackle	114	0.97
California quail	173	11.82
Northern bobwhite	178	5.70
Rock dove	335	5.49
Chukar	578	35.08
Leghorn cockerel	700	21.24
Mallard	1082	79.00

Source: Data from K. R. Solomon et al. "Chlorpyrifos: Ecotoxicological Risk Assessment for Birds and Mammals in Corn Agroecosystems." *Human Ecological Risk Assessment* 7 (2001):497–632. With permission from Taylor and Francis.

```
        1 90
        1 114
        1 173
        1 178
        1 335
        1 578
        1 700
        1 1082];
bodyweight = X(:,2);
LD50 = [0.83 0.69 6.15 1.40 0.97 11.82 5.70 5.49 35.08 21.24 79.00]';

[b,bint,r,rint,stats] = regress(LD50,X,0.05)

yhat = b(1)+b(2)*bodyweight;
plot(bodyweight,LD50,'o')
hold on
plot(bodyweight,yhat)
xlabel('Body Weight,g')
ylabel('Observed and Predicted LD50')
```

The output in the Command Window is:

```
b =
   -4.8257
    0.0649

bint =
  -13.6444    3.9929
    0.0452    0.0846

r =
    3.8384
    2.1407
    5.6536
    0.3844
   -1.6033
    5.4173
   -1.0272
  -11.4272
    2.3911
  -19.3672
   13.5994

rint =
  -16.4039   24.0807
  -18.3999   22.6813
  -14.6142   25.9214
  -20.4172   21.1859
  -22.4726   19.2659
  -15.2467   26.0814
  -22.1386   20.0842
  -30.6680    7.8137
  -18.0777   22.8599
  -31.5163   -7.2181
    4.9328   22.2660

stats =
    0.8603   55.4379    0.0000   86.7477
```

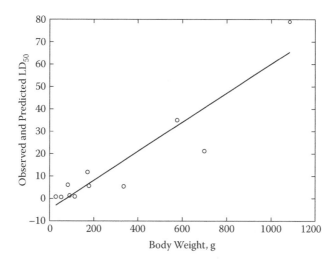

FIGURE 7.2 Observed and predicted LD$_{50}$ values as a function of body weight in birds.

The parameter estimates were the intercept, $b(1) = -4.8257$, the first element in the vector **b**, and $b(2) = 0.0649$, the slope or the second element in vector **b**. The vector **stats** gives a value of R^2 of 0.8603, an F value of 55.4379, and a p value of 0.0000. In MATLAB, the default numeric format prints four decimal places. This gives a p value of 0.0000. Although this would indicate a pretty good fit, a more precise value may be more appropriate. The numeric format can be changed to a floating-point format by typing '**format short e**' in the Command Window, and then typing **stats** to get the new values of R^2, F, p, and the error variance estimate. The p value now is seen as 3.9100e − 005. To return to the default format, type **format short** in the Command Window. The graph of the observations and the regression line is shown in Figure 7.2).

7.1.2 FUNCTION: POLYFIT

The MATLAB function, **polyfit**, can be used to fit a polynomial model. The syntax for using **polyfit** for polynomial regression is:

$$p = \mathtt{polyfit(x,y,n)},$$

or

$$[\mathtt{p,S}] = \mathtt{polyfit(x,y,n)}$$

The syntax **p = polyfit(x,y,n)** finds the coefficients of a polynomial $p(x)$ of degree n that best fits the data y. The result p is a row vector of length $n + 1$ containing the polynomial coefficients in descending powers:

$$y = p_{(1)}x^n + p_{(2)}x^{n-1} + \ldots + p_{(n)}x + p_{(n+1)} \tag{7.4}$$

The syntax **[p,S] = polyfit(x,y,n)** returns the polynomial coefficients p and a structure S for use with **polyval** to obtain error estimates for predictions.

The syntax **[f,delta] = polyval(p,x,S)** returns the value of a polynomial of degree **n** evaluated at **x**. The input argument p is a vector of length **n+1** whose elements are the coefficients obtained from **polyfit**. **delta** is an estimate of the standard deviation of the error in predicting a future observation at **x** by p(**x**). **polyval** evaluates **p** at each element of **x**. If the errors in the data **y** are independent normal with constant variance, **polyval** produces error bounds **y±delta** that contain at least 50% of the predictions of future observations at **x**.

TABLE 7.2
**Mercury Concentrations in ppb in *Daphnia pulex*
during a 14-Day Elimination Experiment**

Replicate	Day							
	0	**1**	**2**	**3**	**7**	**9**	**11**	**14**
1	9699	4616	4170	2332	1567	1313	557	588
2	11060	5281	5169	2868	1570	827	529	648
3	9458	5624	5566	4199	2299	1014	531	4361

Source: Data from Dixon, K. R. "Thermal Plumes and Mercury Dynamics in Zooplankton." In *International Conference on Heavy Metals in the Environment, Vol. 2*, 875–886 (Toronto, Ontario, Canada, 1976).

Example 7.2

In this example of the use of **polyfit**, we first plot the data in Table 7.2 for mercury concentration in ppb in *Daphnia pulex* during a 14-day elimination experiment (Huckabee et al. 1977, Dixon et al. 1977). There are three replicates per day. The results show a classic one-compartment, negative-exponential, elimination relationship (Figure 7.3).

In this example, we want to linearize the data using a log transformation and then use a single-term polynomial regression to fit to the data. To check whether a log transformation is adequate for a simple linear regression, we first plot the transformed data (Figure 7.4). Visual inspection tells us that simple linear regression should provide a good fit to the data. We now can proceed with fitting a linear model using polynomial regression of degree 1. The MATLAB m-file is **polylinregr**:

```
%m-file polylinregr
%program to do linear regression using the MATLAB function polyfit

conc = [9699  4616  4170  2332  1567  1313  557  588 ...
       11060  5281  5169  2868  1570   827  529  648 ...
        9458  5624  5566  4199  2299  1014  531  436];
```

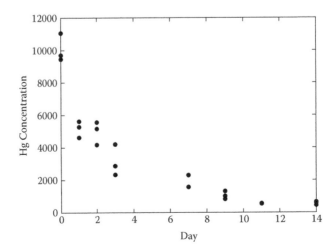

FIGURE 7.3   Plot of data in Table 7.2.

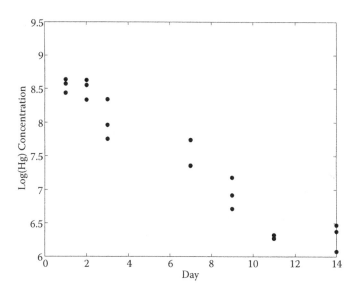

FIGURE 7.4 Plot of log transformed Hg concentration in Table 7.2.

```
allday = [0 1 2 3 7 9 11 14 ...
          0 1 2 3 7 9 11 14 ...
          0 1 2 3 7 9 11 14];
day = [0 1 2 3 7 9 11 14];

logconc = log(conc);
plot(allday,conc,'o')
xlabel('Day')
ylabel('Hg Concentration')

figure(2)
plot(allday,logconc,'o')
xlabel('Day')
ylabel('log(Hg) Concentration')

[p,S] = polyfit(allday, logconc, 1)
[f,delta] = polyval(p,allday,S)
table = [allday logconc f delta]

figure(3)
plot(allday,logconc,'o')
hold on
plot(allday,f)
xlabel('Day')
ylabel('Observed and Predicted Log(Hg Concentration)')
```

Note that the days must be replicated in the matrix **allday** to match the dimensions of the Hg concentration matrix, **conc**. This m-file generates Figures 7.3, 7.4, and 7.5 as well as the regression statistics. The output from the m-file includes: (1) the coefficients in the vector *p* (the slope and the intercept), (2) the parameters in *S*, (3) the predicted values, *f*, in columns 1–8, 9–16, and 17–24; the three sets of columns are identical because the predicted value is the same for each replicate of logconc, (4) the standard deviations in **delta**; again the columns are replicated, and (5) a table with the *x* values, allday, the observed *y* values, allconc, the predicted values *f*, and delta. Only *p*, *f*, and delta are listed below.

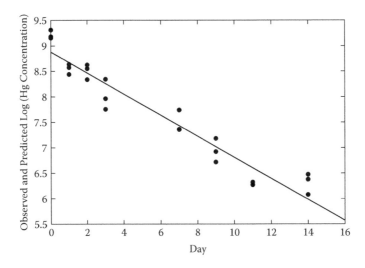

FIGURE 7.5   Plot of regression line with log transformed Hg concentration data from Table 7.2.

```
p =
  -0.2067     8.8781


f =
  Columns 1 through 8
    8.8781   8.6714   8.4647   8.2580   7.4312   7.0179   6.6045   5.9844
  Columns 9 through 16
    8.8781   8.6714   8.4647   8.2580   7.4312   7.0179   6.6045   5.9844
  Columns 17 through 24
    8.8781   8.6714   8.4647   8.2580   7.4312   7.0179   6.6045   5.9844


delta =
  Columns 1 through 8
    0.2966   0.2939   0.2918   0.2901   0.2884   0.2905   0.2946   0.3041
  Columns 9 through 16
    0.2966   0.2939   0.2918   0.2901   0.2884   0.2905   0.2946   0.3041
  Columns 17 through 24
    0.2966   0.2939   0.2918   0.2901   0.2884   0.2905   0.2946   0.3041
```

The regression line now can be plotted with the observed data (Figure 7.5).

7.1.3   FUNCTION: REGSTATS

The MATLAB function **regstats** provides regression diagnostics for linear regression models. As with the MATLAB function **regress**, it is designed for linear multiple regression (more than one independent variable) but works equally well with simple linear regression models (one independent variable). It has the advantage over the function **regress** in that a column of ones is not required in the independent variable matrix. It also provides a greater number of statistics for model testing and comparison. The syntax for **regstats** is:

regstats(responses,data,model)

The **regstats** function regresses measurements in the vector **responses** on values in the matrix, **data**, using a linear multiple regression model. The parameter, **model**, controls the order of the regression model. By default, **regstats** uses a linear additive model with a constant term. The input, **model**, is optional and can be any of the following strings:

```
'linear'              Constant and linear terms (the default)
'interaction'         Constant, linear, and interaction terms
'quadratic'           Constant, linear, interaction, and squared terms
'purequadratic'       Constant, linear, and squared terms
```

The syntax **stats = regstats(responses,data,model,whichstats)** creates an output structure, **stats**, containing the statistics listed in **whichstats**. **whichstats** can be a single string, such as **'fstat'**, or a cell array of strings such as **{'yhat' 'mse' 'rsquare'}**. By default, **regstats** returns all statistics. Some of the valid statistic strings are:

```
Name                  Meaning
'Q'                   Q from the QR Decomposition of the design matrix
'R'                   R from the QR Decomposition of the design matrix
'beta'                Regression coefficients
'covb'                Covariance of regression coefficients
'yhat'                Fitted values of the response data
'r'                   Residuals
'mse'                 Mean squared error
'rsquare'             R-square statistic
'adjrsquare'          Adjusted R-square statistic
'leverage'            Leverage
'hatmat'              Hat (projection) matrix
's2_i'                Delete-1 variance
'beta_i'              Delete-1 coefficients
'standres'            Standardized residuals
'studres'             Studentized residuals
'dfbetas'             Scaled change in regression coefficients
'dffit'               Change in fitted values
'dffits'              Scaled change in fitted values
'covratio'            Change in covariance
'cookd'               Cook's distance
'tstat'               t statistics for coefficients
'fstat'               F statistic
'dwstat'              Durbin Watson statistic
'all'                 Create all of the above statistics
```

To obtain individual statistics in the Command Window, type **stats.'name'** in the Command Window.

Example 7.3

This example uses the MATLAB function, **regstats**, to fit a linear regression model to the log transformed data in Example 7.1, generate diagnostic statistics, and plot residuals versus fitted values. The MATLAB m-file is **regressstats**:

```
%m-file regressstats
%program to do linear regression using the MATLAB function regstats

bodyweight = [28 52 82 90 114 173 178 335 578 700 1082];
LD50 = [0.83 0.69 6.15 1.40 0.97 11.82 5.70 5.49 35.08 21.24 79.00]';
```

```
stats = regstats(LD50, bodyweight, 'linear', {'all'})
scatter(stats.yhat,stats.r)
hold on
a=[-20:1:80];
x = zeros(1,length(a));
plot(a,x,'--')
xlabel('Fitted Values'); ylabel('Residuals');

figure
plot(bodyweight,LD50,'o')
hold on
plot(bodyweight,stats.yhat)
xlabel('Bodyweight')
ylabel('Observed and Predicted LD_{50}')
```

The statistics generated are output in the Command Window:

```
stats =
        source: 'regstats'
             Q: [11x2 double]
             R: [2x2 double]
          beta: [2x1 double]
          covb: [2x2 double]
          yhat: [11x1 double]
             r: [11x1 double]
           mse: 86.7477
       rsquare: 0.8603
    adjrsquare: 0.8448
      leverage: [11x1 double]
        hatmat: [11x11 double]
          s2_i: [11x1 double]
        beta_i: [2x11 double]
      standres: [11x1 double]
       studres: [11x1 double]
       dfbetas: [2x11 double]
         dffit: [11x1 double]
        dffits: [11x1 double]
      covratio: [11x1 double]
         cookd: [11x1 double]
         tstat: [1x1 struct]
         fstat: [1x1 struct]
        dwstat: [1x1 struct]
```

The three scalar statistics: **mse**, **rsquare**, and **adjrsquare**, are ouput directly in the **stats** output structure. To retrieve other statistics, the cell array statistic is typed in the Command Window. For example, to retrieve the regression coefficients, **beta**, F statistic, **fstat**, the residuals, **r**, and the predicted values, **yhat**, the following statements are entered, followed by the answer:

```
>> stats.beta
ans =
   -4.8257
    0.0649

>> stats.fstat
ans =
    sse: 780.7297
    dfe: 9
```

```
    dfr: 1
    ssr: 4.8091e+003
      f: 55.4379
   pval: 3.9100e-005

>> stats.r
ans =
    3.8384
    2.1407
    5.6536
    0.3844
   -1.6033
    5.4173
   -1.0272
  -11.4272
    2.3911
  -19.3672
   13.5994

>> stats.yhat
ans =
   -3.0084
   -1.4507
    0.4964
    1.0156
    2.5733
    6.4027
    6.7272
   16.9172
   32.6889
   40.6072
   65.4006
```

Note that the values of these statistics are comparable to those obtained by using the **regress** function in Example 7.1. The plot of the residuals over the fitted LD_{50} values is shown in Figure 7.6 using the scatter plot statement: **scatter(stats.yhat,stats.r)**.

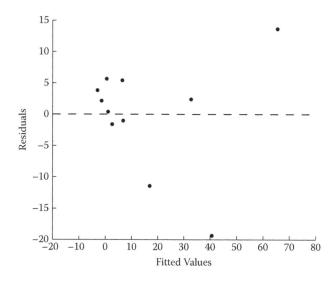

FIGURE 7.6   Plot of residuals over fitted LD_{50} values using data in Table 7.1.

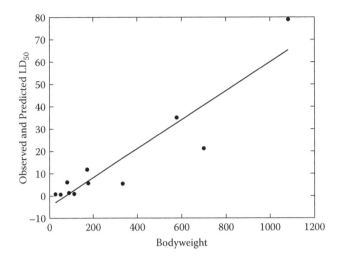

FIGURE 7.7   Observed and predicted LD_{50} values (Table 7.1).

The plot shows little bias, as the residuals do not show a trend. They do show, however, that the fit is relatively poor at higher LD_{50} values. The second plot (Figure 7.7) shows the same linear relationship as the regression in Example 7.1.

7.2   NONLINEAR REGRESSION

Nonlinear regression is used when the relationship is nonlinear in the parameters. Nonlinear regression algorithms often require initial estimates of the parameters. Methods of obtaining initial parameter estimates include (1) graphing, (2) guessing, and (3) linear regression using a linearized model. Nonlinear regression is an iterative procedure and may find local minima. It should be obvious if this happens because the model will not fit the data. The algorithm is then modified, such as changing tolerance values or initial parameter values, until an adequate fit is obtained.

7.2.1   FUNCTION: NLINFIT

The Statistics Toolbox provides the function **nlinfit** for finding parameter estimates in nonlinear modeling. Nonlinear least-squares regression using **nlinfit** fits a nonlinear model to data by the Gauss-Newton method. The syntax for **nlinfit** is:

```
[beta,r,J,COVB,mse] = nlinfit(x,y,@fun,beta0)
```

The function **nlinfit** returns the least squares parameter estimates **beta**, the residuals **r**, the Jacobian matrix **J**, the estimated covariance matrix **COVB** for the fitted coefficients, and an estimated **mse** of the variance of the error term, the mean squared error. Typically, **x** is a vector of predictor (independent variable) values and **y** is a vector of response (dependent variable) values. The **@fun** function takes the array of input data, **x**, and the initial parameter estimates, **beta0**, and returns a vector of the predicted responses, **yhat**. You can use these outputs with the MATLAB functions **nlpredci** to produce error estimates on predictions, and with **nlparci** to produce error estimates on the estimated coefficients.

The function **nlpredci** generates nonlinear least squares prediction confidence intervals (nlpredci). The syntax is:

```
[ypred,delta] = nlpredci(@fun,x,beta,r,'covar',sigma).
```

The function **nlpredci** actually returns predictions (**ypred**) and 95% confidence interval half-widths (**delta**) for the function **@fun** at input values **x**. The function **@fun** inputs two variables: the coefficient vector, **beta**, and the vector of predictor (independent variable) values, **x**, and returns a vector of fitted **y** values, **ypred**.

The function **nlparci** returns 95% nonlinear least squares parameter confidence intervals **betaci** for the parameters in **beta**. The syntax is:

```
betaci = nlparci(beta,r,'covar',sigma)
```

Both **nlparci** and **nlpredci** depend upon **nlinfit** for input. Therefore, before using either **nlparci** or **nlpredci**, one must use **nlinfit** to get estimated coefficient values **beta**, residuals, **r**, and estimated coefficient covariance matrix **sigma**.

Example 7.4

This example fits a nonlinear model to a set of data (Table 7.3) on the cumulative mortality over 21 days in *Daphnia magna* during exposure to chlorpyrifos (Naddy et al. 2000). Following the steps in regression described previously, we first plot the data (Figure 7.8). The graph suggests a typical sigmoid curve (see Chapter 2, Table 2.1 and Chapter 5, Figure 5.4). The second step in regression is to select a model to fit to the data. We chose the logistic function in the form, reproduced in Equation (7.5), in which the initial parameter values can be visually estimated.

$$F(t) = \frac{P_1}{1 + e^{(2.2/P_3)(P_2 - t)}} \tag{7.5}$$

TABLE 7.3
Cumulative Mortality in *Daphnia magna* over 21 Days following Exposure to Chlorpyrifos

Day	Cumulative Mortality	Day	Cumulative Mortality
0.042	0	8.0	0
0.125	0	9.0	1
0.25	0	10.0	6
0.5	0	11.0	6
1.0	0	12.0	10
1.5	0	13.0	15
2.0	0	14.0	20
2.5	0	15.0	20
3.0	0	16.0	20
3.5	0	17.0	20
4.0	0	18.0	20
5.0	0	19.0	20
6.0	0	20.0	20
7.0	0	21.0	20

Source: Data from R. B. K. Naddy, A. Johnson, and S. J. Klaine. 2000. "Response of *Daphnia magna* to Pulsed Exposures of Chlorpyrifos." *Environmental Toxicology and Chemistry* 19:423–443. With permission from John Wiley & Sons.

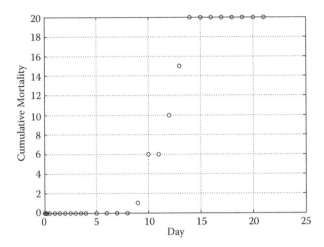

FIGURE 7.8 Cumulative mortality in *Daphnia magna* during a 21-day experiment following exposure to chlorpyrifos. (Data from R. B. Naddy, K. A. Johnson, and S. J. Klaine. 2000. "Response of *Daphnia magna* to Pulsed Exposures of Chlorpyrifos." *Environmental Toxicology and Chemistry* 19, 423–431. With permission from John Wiley & Sons.)

where

P_1 = maximum mortality response
P_2 = day of 50% cumulative mortality
P_3 = number of days between 10% and 50% cumulative mortality

We could estimate all three parameters, P_1, P_2, and P_3; however, we know that P_1 has to be 20.0 because there were only twenty *Daphnia* used in the experiment. That leaves only P_2 and P_3, which we estimated from the graph as 12 and 3, respectively. The **fun** function used in **nlinfit** we call **mort**. This function includes the model to be fitted, along with the names of the initial parameter estimates in the vector **beta**.

The m-file for **nlinfit** is **daphniachlor**:

```
% m-file daphniachlor
% program to fit nonlinear model to mortality data
% (source: Naddy, et al. 2000)
load C:\MATLAB\chlorpyrifos_data.rtf

day=chlorpyrifos_data(:,1);
mortppb12=chlorpyrifos_data(:,2);
plot(day,mortppb12, 'ko')
xlabel('Day')
ylabel('Cumulative Mortality');

beta0 = [12; 3];          %Initial parameter estimates
newx = 0:.1:25;
%Generate parameter values, residuals, and Jacobian
[beta,r,J,sigma] = nlinfit(day,mortppb12,@mort,beta0)

%Confidence intervals on parameters
betaci = nlparci(beta,r,'covar',sigma)

%Confidence intervals on predicted values
%Delta is half widths of 95% CI

[yhat, delta] = nlpredci(@mort,newx,beta,r,'covar',sigma);
```

```
ucl = yhat + delta';
lcl = yhat - delta';

figure
plot(day, mortppb12,'ko')
xlabel('Day')
ylabel('Cumulative Mortality');
hold on
plot(newx, yhat,'r-')
plot(newx, ucl, 'b-')
plot(newx, lcl, 'g-')
hold off
```

The **@fun** function **mort** is:

```
function yhat = mort(beta,day)

p2 = beta(1);
p3 = beta(2);

yhat = 20./(1+exp((2.2/p3)*(p2-day)));
```

The output listed in the Command Window includes the parameter estimates **beta**, the residuals **r**, the Jacobian matrix **J**, the estimated coefficient covariance matrix, **sigma**, and the confidence intervals for the parameter estimates **betaci**.

```
beta =
    11.7195
     2.3147


r =
    -0.0003
    -0.0003
    -0.0004
    -0.0005
    -0.0008
    -0.0012
    -0.0019
    -0.0031
    -0.0050
    -0.0081
    -0.0130
    -0.0336
    -0.0867
    -0.2229
    -0.5665
    -0.4025
     2.7353
    -0.7078
    -1.3250
    -0.4307
     2.0542
     0.8475
     0.3364
     0.1314
     0.0510
     0.0197
     0.0076
     0.0030
```

```
J =
   -0.0003      0.0015
   -0.0003      0.0016
   -0.0004      0.0017
   -0.0004      0.0022
   -0.0007      0.0033
   -0.0011      0.0051
   -0.0018      0.0078
   -0.0030      0.0118
   -0.0048      0.0180
   -0.0077      0.0273
   -0.0124      0.0412
   -0.0319      0.0926
   -0.0821      0.2028
   -0.2094      0.4271
   -0.5231      0.8407
   -1.2394      1.4563
   -2.5964      1.9289
   -4.2373      1.3172
   -4.6690     -0.5658
   -3.3507     -1.8537
   -1.7519     -1.7260
   -0.7713     -1.0932
   -0.3143     -0.5812
   -0.1240     -0.2830
   -0.0483     -0.1311
   -0.0187     -0.0589
   -0.0073     -0.0259
   -0.0028     -0.0112


sigma =
    0.0094     -0.0000
   -0.0000      0.0354


betaci =
   11.5199     11.9192
    1.9280      2.7013
```

The resulting nonlinear regression fit and 95% confidence intervals are shown in Figure 7.9. The solid line is the fitted mortality. The dashed lines are lower and upper 95% confidence limits.

Example 7.5

This example takes the same elimination data from Huckabee et al. (1976) and Dixon (1976) as in Example 7.2 and fits a negative exponential model to the data using nonlinear least squares regression instead of transforming the data to use linear regression. As in that example, a plot of the data confirms a nonlinear relationship between mercury concentration and time (Figure 7.10). We use the negative exponential model described in Chapter 2 (Table 2.1, Equation [2.3] and Example 2.2) and Chapter 5 (Example 5.4).

We use **nlinfit** to obtain the fitted parameter values, **beta**, and residuals, **r**. All of the output from **nlinfit**, **nlpredci**, and **nlparci** is output to the Command Window and is generated by the m-file **daphnia _ mercury2**:

```
%m-file daphnia_mercury2
%program to fit exponential model with nonlinear least squares
%regression using nlinfit

conc = [9699  4616  4170  2332  1567  1313   557   588
```

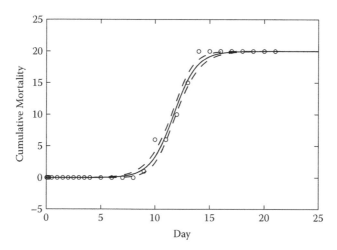

FIGURE 7.9 Results of the fit of a logistic model to cumulative mortality. Solid line is the predicted mortality. Dashed lines are the lower and upper 95% confidence limits, respectively. (Data from R. B. Naddy, K. A. Johnson, and S. J. Klaine. 2000. "Response of *Daphnia magna* to Pulsed Exposures of Chlorpyrifos." *Environmental Toxicology and Chemistry* 19, 423–431. With permission from John Wiley & Sons.)

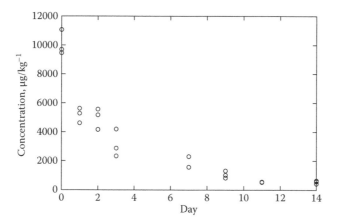

FIGURE 7.10 Hg concentrations from a 14-day elimination study in *Daphnia* spp.

```
          11060   5281   5169   2868   1570    827   529   648
           9458   5624   5566   4199   2299   1014   531   436];

allconc = [9699   4616   4170   2332   1567   1313   557   588 ...
          11060   5281   5169   2868   1570    827   529   648 ...
           9458   5624   5566   4199   2299   1014   531   436];
allday = [0 1 2 3 7 9 11 14 ...
          0 1 2 3 7 9 11 14 ...
          0 1 2 3 7 9 11 14];
day = [0 1 2 3 7 9 11 14];

meanconc = mean(conc)
stdconc = std(conc)

beta0 = [8900; .3144; 400];
newx = 0:.1:15;
```

```
[beta,r,J,sigma] = nlinfit(allday,allconc,@daphnia_mercury,beta0)

betaci = nlparci(beta,r,'covar',sigma)

[yhat, delta] = nlpredci(@daphnia_mercury,newx,beta,r,'covar',sigma);

ucl = yhat + delta;
lcl = yhat - delta;

plot(allday, allconc,'ko')
xlabel('Day')
ylabel('Concentration, \mug\cdotkg^{-1}')

figure
errorbar(day,meanconc,stdconc,'k+')
hold on
plot(newx, yhat,'k-')
plot(newx, ucl, 'k--')
plot(newx, lcl, 'k--')
xlabel('Day')
ylabel('Hg Concentration, \mug\cdotkg^{-1}')
```

The **@fun** function, **daphnia _ mercury**, with the parameter names and exponential elimination model is:

```
function yhat = daphnia(beta,day)

b1 = beta(1);
b2 = beta(2);
b3 = beta(3);

yhat = (b1-b3)*exp(-b2*day)+b3;
```

The output in the Command Window is:

```
meanconc =
  1.0e+004 *
    1.0072   0.5174   0.4968   0.3133   0.1812   0.1051   0.0539   0.0557

stdconc =
  863.7907   512.5001   719.3082   961.2965   421.7570   245.1415   15.6205
109.2764

beta =
  1.0e+003 *
     9.6384
     0.0005
     0.8643

r =
  1.0e+003 *
  Columns 1 through 8
    0.0606   -1.7438   -0.1362   -0.6881   0.3710   0.3186   -0.3583   -0.2888
  Columns 9 through 16
    1.4216   -1.0788   0.8628   -0.1521   0.3740   -0.1674   -0.3863   -0.2288
  Columns 17 through 24
   -0.1804   -0.7358   1.2598   1.1789   1.1030   0.0196   -0.3843   -0.4408
```

```
J =
  1.0e+003 *
    0.0010         0         0
    0.0006   -5.4956    0.0004
    0.0004   -6.8842    0.0006
    0.0002   -6.4677    0.0008
    0.0000   -2.3225    0.0010
    0.0000   -1.1715    0.0010
    0.0000   -0.5617    0.0010
    0.0000   -0.1757    0.0010
    0.0010         0         0
    0.0006   -5.4956    0.0004
    0.0004   -6.8842    0.0006
    0.0002   -6.4677    0.0008
    0.0000   -2.3225    0.0010
    0.0000   -1.1715    0.0010
    0.0000   -0.5617    0.0010
    0.0000   -0.1757    0.0010
    0.0010         0         0
    0.0006   -5.4956    0.0004
    0.0004   -6.8842    0.0006
    0.0002   -6.4677    0.0008
    0.0000   -2.3225    0.0010
    0.0000   -1.1715    0.0010
    0.0000   -0.5617    0.0010
    0.0000   -0.1757    0.0010

sigma =
  1.0e+005 *
    1.9262    0.0001    0.1407
    0.0001    0.0000    0.0001
    0.1407    0.0001    0.6919

betaci =
  1.0e+004 *
    0.8726    1.0551
    0.0000    0.0001
    0.0317    0.1411
```

We used the format **short** **e** to get more precise estimates of the elimination rate constant **betahat(2)** and the confidence interval on the rate constant **betaci(2,:)**.

```
  betahat(2)

ans =
  4.6788e-001

  betaci(2,:)

ans =
  3.4248e-001   5.9328e-001
```

The fit of the model to the data and the 95% confidence intervals for the data are shown in Figure 7.11.

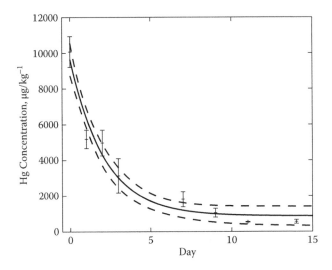

FIGURE 7.11   Observed and fitted Hg concentrations. Error bars are 95% CI on the mean observed data. Continuous line is the mean and dashed lines the 95% CI on fitted model.

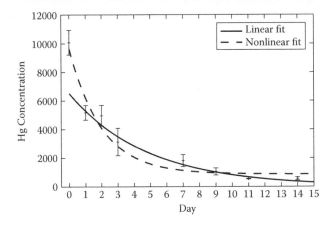

FIGURE 7.12   Plots of linear and nonlinear models to mercury elimination data in Table 7.2.

7.3   COMPARISON BETWEEN LINEAR AND NONLINEAR REGRESSIONS

We conducted examples of the two ways of fitting the negative exponential model to a nonlinear relationship: (1) transforming the data to linearize the relationship and then using linear regression (Example 7.2) and (2) using nonlinear regression (Example 7.5). How do the results from these two approaches compare? That is, which gives a better fit to the data? And, how different are the parameter estimates?

There is considerable difference in the fit of the two regression methods (Figure 7.12). Although the linear regression does a reasonable job of fitting a linear model to the data from days 1 to 14, it greatly underestimates the initial mercury concentration on day zero. And, as is shown in Table 7.4 the elimination rate is greatly underestimated. This illustrates the bias inherent in using a linear model to fit a nonlinear relationship.

TABLE 7.4
Parameter Values as Estimated
by Linear and Nonlinear Regression

	Linear	Nonlinear
Maximum Y value	6509.4	9638.4
Minimum Y value	Not estimated	864.3
Elimination rate (Slope)	−0.2067	−0.50

TABLE 7.5
Deer Body Weights at Ages 0–7.5

Age	0	0.5	1.5	2.5	3.5	4.5	7.5
Body Weight	0	78.2	126.0	150.0	214.5	236.0	237.0

EXERCISES

This exercise uses data on the age-specific body weights in deer.

1. Write an m-file to complete the following steps and answer the questions.
2. Plot deer body weight over age.
3. Is the relationship linear or nonlinear?
4. Select a model to fit the data.
5. Linearize the model and plot the linearized data and model.
6. Use linear regression to fit the linearized model to the linearized data.
7. Use nonlinear regression to fit the nonlinear model to the data in Table 7.5, including 95% confidence intervals.
8. Plot the predicted values and the upper and lower confidence limits.
9. Plot the data, the predicted body weights using parameters from linear regression, and the predicted body weights using the parameters from the nonlinear regression.
10. Compare the two plots of the predicted values and the data.
11. Why do the predictions based upon the linearized model do a better job of predicting body weight than the comparison in Section 7.3?

REFERENCES

Abramowitz, M., and I. Stegun, eds. 1964. *Handbook of Mathematical Functions with Formulas, Graphs and Mathematical Tables*. Washington, DC: U.S. Government Printing Office.
Berkson J. 1950. "Are There Two Regressions?" *Journal of the American Statistical Association* 45:164–180.
Carroll, R. J., and D. Ruppert. 1988. *Transformation and Weighting in Regression*. New York: Chapman & Hall.
Chatterjee, S., and A. S. Hadi. 1986. "Influential Observations, High Leverage Points, and Outliers in Linear Regression." *Statistical Science* 1:379–416.
Cleveland, W. S. 1979. "Robust Locally Weighted Regression and Smoothing Scatterplots." *Journal of the American Statistical Association* 74:829–836.
Cleveland, W. S., and S. J. Devlin. 1988. "Locally Weighted Regression: An Approach to Regression Analysis by Local Fitting." *Journal of the American Statistical Association* 83:596–610.
Dixon, K. R. 1977. "Thermal Plumes and Mercury Dynamics in Zooplankton." In *International Conference on Heavy Metals in the Environment, Vol. 2*, 875–886. Toronto, Ontario, Canada.
Fuller, W. A. 1987. *Measurement Error Models*. New York: Wiley.
Graybill, F. A. 1976. *Theory and Application of the Linear Model*. North Sciutate, MA: Duxbury Press.

Graybill, F. A., and H. K. Iyer. 1994. *Regression Analysis: Concepts and Applications*. Belmont, CA: Duxbury Press.

Harter, H. L. 1983. "Least Squares." In *Encyclopedia of Statistical Sciences*, ed. S. Kotz and N. L. Johnson, 593–598. New York: Wiley.

Huckabee, J. W., R. A. Goldstein, S. A. Janzen, and S. E. Woock. 1977. "Methylmercury in a Freshwater Foodchain." In *International Conference on Heavy Metals in the Environment, Vol. 2, Part 1,* 199–216. Toronto, Ontario, Canada.

Montgomery, D.C. 2001. *Design and Analysis of Experiments, 5th Edition*. New York: Wiley.

Naddy, R. B., K. A. Johnson, and S. J. Klaine. 2000. "Response of *Daphnia magna* to Pulsed Exposures of Chlorpyrifos." *Environmental Toxicology and Chemistry* 19:423–431.

Neter, J., W. Wasserman, and M. Kutner. 1983. *Applied Linear Regression Models*. Homewood, IL: Richard D. Irwin, Inc.

Ryan, T. P. 1997. *Modern Regression Methods*. New York: Wiley.

Seber, G. A. F., and C. F. Wild. 1989. *Nonlinear Regression*. New York: Wiley.

Solomon, K. R., J. P. Giesy, R. J. Kendall, L. B. Best, J. R. Coats, K. R. Dixon, M. J. Hooper, E. E. Kenaga, and S. T. McMurry. 2001. "Chlorpyrifos: Ecotoxicological Risk Assessment for Birds and Mammals in Corn Agroecosystems." *Human Ecological Risk Assessment* 7:497–632.

Stigler, S. M. 1978. "Mathematical Statistics in the Early States." *The Annals of Statistics* 6:239–265.

Stigler, S. M. 1986. *The History of Statistics: The Measurement of Uncertainty before 1900*. Cambridge, MA: The Belknap Press of Harvard University Press.

Watt, K. E. F. 1968. *Ecology and Resource Management: A Quantitative Approach*. New York: McGraw-Hill.

8 Designing Simulation Experiments

Experimental design of simulation experiments is virtually identical to the design of experiments for real systems, whether laboratory experiments or field studies. The main difference is that with simulation experiments, the experimenter has complete control over the experiment. He or she can control all the variables in the experiment, whereas in real-world experiments there are always factors that cannot be controlled by the experimenter. In simulation experiments, variables represented by input parameters are called *factors* whereas the resulting dependent variable outcomes are called *responses*. Experimental designs usually are developed after parameter estimation has been completed so that the appropriate parameters and their levels can be incorporated into the design. Usually, not all parameters in a model would be included as factors. Only those factors we suspect will have the greatest effect on the response, or are of particular interest, need to be included. Likewise, there may be several response variables of interest. The particular design is constructed to maximize the amount of information with a minimum amount of effort (Martin 1968). Therefore, we are interested in designing *efficient* simulation experiments, i.e., determining the fewest number of experiments that will provide the necessary data for statistical analysis.

The particular design used in a simulation experiment depends upon the purpose of the experiment (Hunter and Naylor 1970). As in experimental statistics in general, simulation experiments can be divided into those designed to test hypotheses and those designed to make predictions. Testing hypotheses of differences in response variable means, as a function of different factors, usually requires some kind of factorial design. Prediction, using least squares regression models requires a design in which the ranges of values of all the factors are set in such a way as to enclose the values of the response variable. Recall that making predictions outside the range of the data is risky. This type of simulation experiment is designed to generate a response surface model.

8.1 FACTORIAL DESIGNS

Factorial designs include both *full factorial designs* and *fractional factorial designs*. In full factorial designs, we run experiments in all possible combinations of factors. In fractional factorial designs, experiments are run only on a "fraction" of the full factorial design. These fractional designs often are used to screen the factors to decide which are the most important. The two types of designs are explored more fully in the following sections with MATLAB procedures for determining the combination of factor levels in each experiment.

8.1.1 FULL FACTORIAL DESIGNS

In a full factorial experiment with two levels of each factor (parameter), the total number of simulation runs is 2 raised to the number of variables. For example, a model with 5 parameters will have $2^5 = 32$ runs; a model with 8 parameters will have $2^8 = 256$ runs; and a model with 12 parameters will have $2^{12} = 4096$ runs. Obviously, in a simulation experiment, we would like to limit the number of parameters to those having the greatest effect on the output or those for which we have some special interest. Before we consider ways to reduce the number of experiments, we review the design that includes all pertinent variables and their interactions, the full factorial design.

For example, consider an experiment with five variables with 2 levels each. As we saw previously, a full factorial will require 32 runs. The MATLAB function **ff2n(n)** generates two-level full-factorial designs. N is the number of columns representing the number of variables or parameters. The number of rows is 2^N. To generate the design as an array, we can use the command **x=ff2n(n)**. To generate the design for our example of $2^5 = 32$ runs, we type the command, **x=ff2n(5)** in the Command Window:

```
>> x=ff2n(5)
```

The resulting array appears in the Command Window:

```
x =
```

0	0	0	0	0
0	0	0	0	1
0	0	0	1	0
0	0	0	1	1
0	0	1	0	0
0	0	1	0	1
0	0	1	1	0
0	0	1	1	1
0	1	0	0	0
0	1	0	0	1
0	1	0	1	0
0	1	0	1	1
0	1	1	0	0
0	1	1	0	1
0	1	1	1	0
0	1	1	1	1
1	0	0	0	0
1	0	0	0	1
1	0	0	1	0
1	0	0	1	1
1	0	1	0	0
1	0	1	0	1
1	0	1	1	0
1	0	1	1	1
1	1	0	0	0
1	1	0	0	1
1	1	0	1	0
1	1	0	1	1
1	1	1	0	0
1	1	1	0	1
1	1	1	1	0
1	1	1	1	1

In this design, there are five variables, each represented by one of the five columns. Each row represents a simulation run with a 0 indicating a low value of the variable and a 1 indicating a high value. The first run would have low values for all five variables. The second run would have low values for the first four variables and the high value for the fifth variable, and so on.

TABLE 8.1
Fractional Factorial Experimental Designs

Number of Variables	Number of Runs	Degree of Fractionation	Type of Design	Method of Introducing "New" Factors	Blocking (with no main effect or interaction confounded)	Method of Introducing Blocks
5	16	½	2_V^{5-1}	±5 = 1234	not available	
6	32	½	2_{VI}^{6-1}	±6 = 12345	2 blocks of 16 runs	$B_1 = 123$
7	64	½	2_{VII}^{7-1}	±7 = 123456	8 blocks of 8 runs	$B_1 = 1357$ $B_2 = 1256$ $B_3 = 1234$
8	64	¼	2_V^{8-2}	±7 = 1234 ±8 = 1256	4 blocks of 16 runs	$B_1 = 135$ $B_2 = 348$
9	128	¼	2_{VI}^{9-2}	±8 = 13467 ±9 = 23567	8 blocks of 16 runs	$B_1 = 138$ $B_2 = 129$ $B_3 = 789$
10	128	⅛	2_V^{10-3}	±8 = 1237 ±9 = 2345 ±10 = 1346	8 blocks of 16 runs	$B_1 = 149$ $B_2 = 12(10)$ $B_3 = 89(10)$
11	128	1/16	2_V^{11-4}	±8 = 1237 ±9 = 2345 ±10 = 1346 ±11 = 1234567	8 blocks of 16 runs	$B_1 = 149$ $B_2 = 12(10)$ $B_3 = 89(10)$

Source: Adapted from G. E. P. Box, W. G. Hunter, and J. S. Hunter. 1978. *Statistics for Experimenters: An Introduction to Design, Data Analysis, and Model Building.* With permission from John Wiley & Sons.

8.1.2 FRACTIONAL FACTORIAL

It is possible to reduce the number of simulation runs using a fractional factorial design (Box et al. 1978, 2005). These designs assume that higher order (>2) interactions are negligible. The best designs have resolution V or higher (Table 8.1).

In general, to construct a 2^{k-1} fractional factorial of highest possible resolution:

Write a full factorial for the first *k-1* parameters
Associate the *k*th variable with ± the interaction column 123 … (k − 1)

In fractional factorial designs, the number of levels per parameter usually is limited to 2. A more efficient design that provides for more than 2 levels is *response surface methodology*, which is discussed in the following section.

For example, we first will generate a two-level full factorial design using the fractional factorial function, **fracfact**. The command, **x = fracfact(gen)**, produces the fractional factorial design defined by the generator string **gen**. The generator string, **gen**, must be a sequence of "words" separated by spaces, where a "word" represents a factor or combination of factors. For example, we saw above that the number of experiments in a full factorial design with 5 factors and 2 levels for each factor $N = 2^5$. In general, the generator string consists of *P* words using *K* letters of the alphabet, so that **x** will have $N = 2^K$ rows and *P* columns. In our example then, the number of "words" is 5, but we will use 5 letters. Therefore, this design has 32 rows and 5 columns. To generate the same design as in the previous example, type **x = fracfact('a b c d e')** in the Command Window:

```
>> x = fracfact('a b c d e')

x =

    -1    -1    -1    -1    -1
    -1    -1    -1    -1     1
    -1    -1    -1     1    -1
    -1    -1    -1     1     1
    -1    -1     1    -1    -1
    -1    -1     1    -1     1
    -1    -1     1     1    -1
    -1    -1     1     1     1
    -1     1    -1    -1    -1
    -1     1    -1    -1     1
    -1     1    -1     1    -1
    -1     1    -1     1     1
    -1     1     1    -1    -1
    -1     1     1    -1     1
    -1     1     1     1    -1
    -1     1     1     1     1
     1    -1    -1    -1    -1
     1    -1    -1    -1     1
     1    -1    -1     1    -1
     1    -1    -1     1     1
     1    -1     1    -1    -1
     1    -1     1    -1     1
     1    -1     1     1    -1
     1    -1     1     1     1
     1     1    -1    -1    -1
     1     1    -1    -1     1
     1     1    -1     1    -1
     1     1    -1     1     1
     1     1     1    -1    -1
     1     1     1    -1     1
     1     1     1     1    -1
     1     1     1     1     1

>>>
```

This is the same design as that generated by the full factorial MATLAB function $ff2n(n)$, although in this design, the low parameter value is indicated by −1 and the high value by 1.

In the next part of this example, we generate the fractional factorial design, 2_V^{5-1}, which reduces the number of simulation runs from 32 to 16. The command, $x = fracfact('a\ b\ c\ d\ abcd')$, produces a 16-run fractional factorial design for 5 variables, where the first 4 columns are a 16-run, 2-level full factorial design for the first 4 variables, and the fifth column is the product of the first 4 columns. Therefore, the number of "words" is still 5, but we use only four letters. The fifth column is confounded with the four-way interaction of the first 4 columns.

```
>> x = fracfact('a b c d abcd')

x =

    -1    -1    -1    -1     1
    -1    -1    -1     1    -1
    -1    -1     1    -1    -1
    -1    -1     1     1     1
    -1     1    -1    -1    -1
    -1     1    -1     1     1
    -1     1     1    -1     1
    -1     1     1     1    -1
     1    -1    -1    -1    -1
     1    -1    -1     1     1
     1    -1     1    -1     1
     1    -1     1     1    -1
     1     1    -1    -1     1
     1     1    -1     1    -1
     1     1     1    -1    -1
     1     1     1     1     1
```

Note that the first 4 columns are the same first 4 columns of the previous full factorial example, taking every other row.

The command `[x, conf] = fracfact(gen)` also returns `conf`, a cell array of strings containing the confounding pattern for the design.

8.2   RESPONSE SURFACE DESIGNS

Response surface designs not only can reduce the number of runs compared to a full factorial design, but assure that the fit of the response surface model to the data has a small residual error. Some of the possible designs are presented here, including central composite designs, Box-Behnken designs, and other composite designs.

As with factorial designs, the parameter values are normalized so that the lower parameter value is coded as −1, the higher value as +1, and the median value as 0. For example, assume a parameter that has a low value of 5, a median value of 10, and a high value of 15. The normalized values are calculated as

$$x_1 = (5-10)/5 = -1$$

$$x_2 = (10-10)/5 = 0$$

$$x_3 = (15-10)/5 = 1$$

To normalize the parameters, we subtract the median of the parameter (10) from the parameter value (5, 10, or 15), and then divide by the increment (the difference between the high or low value and the median value (5). Normalization makes the comparison of different designs easier than working with the original values. For a given design, the actual values of a factor are obtained by reversing the normalization procedure.

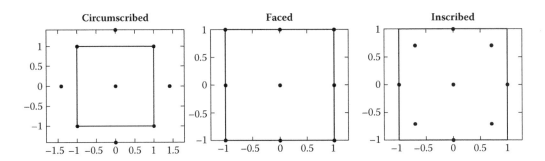

FIGURE 8.1 Central composite designs with two factors.

8.2.1 CENTRAL COMPOSITE DESIGNS

Central composite designs contain three types of experimental runs.

1. A set of center points. These points represent the medians of the values used in the factorial portion. This point is often replicated in order to improve the precision of the experiment.
2. A set of cube (or corner) points. These points are at the corners of the cube representing the high and low values of the factors.
3. A set of axial (or star) points. These points are at the same level as the center points except for one factor, and will take on values both below and above the median of the other factors, and typically both outside the range of the cube points. All factors are varied in this way.

The design can vary according to the relative position of the cube and center points (Figure 8.1). The circumscribed design has the axial points extending beyond the cube points. The faced design has the axial points on the cube face. The inscribed design has the limits of the axial points set to −1 and 1. In other words, the cube is *inscribed* within the axial points. For both circumscribed and faced designs, normalization is based upon the cube points; for the inscribed design, it is based upon the star points.

The MATLAB function **dCC** = **ccdesign(n)** generates a central composite design for **n** factors. The number of factors **n** must be an integer 2 or larger. The output matrix **dCC** is m-by-n, where m is the number of runs in the design. Each row in the matrix represents one run. Each column in the matrix represents a factor. The default design is circumscribed where factor values are normalized on the cube points. For example, a 2-factor design is generated by typing

```
dCC = ccdesign(2)
```

in the Command Window. The resulting output is displayed in the Command Window.

```
dCC =

    -1.0000    -1.0000
    -1.0000     1.0000
     1.0000    -1.0000
     1.0000     1.0000
    -1.4142         0
     1.4142         0
         0    -1.4142
         0     1.4142
         0         0
```

0	0
0	0
0	0
0	0
0	0
0	0
0	0

The first 4 runs use parameter values on the cube; the next 4 runs are on axial points; and the last 8 points are on center points.

An example of a three-factor design is generated by typing the command

```
dCC = ccdesign(3)
```

in the Command Window, which yields the following output:

```
dCC =
```

-1.0000	-1.0000	-1.0000
-1.0000	-1.0000	1.0000
-1.0000	1.0000	-1.0000
-1.0000	1.0000	1.0000
1.0000	-1.0000	-1.0000
1.0000	-1.0000	1.0000
1.0000	1.0000	-1.0000
1.0000	1.0000	1.0000
-1.6818	0	0
1.6818	0	0
0	-1.6818	0
0	1.6818	0
0	0	-1.6818
0	0	1.6818
0	0	0
0	0	0
0	0	0
0	0	0
0	0	0
0	0	0
0	0	0
0	0	0

This design can be visualized by plotting the design points (Figure 8.2).

MATLAB gives you the option of specifying the number of center points, the fraction of the full factorial for the cube portion, and the type of design (whether inscribed, circumscribed, or faced). The syntax for this option is:

```
dCC=ccdesign(nfactors,'pname1',pvalue1,'pname2',pvalue2,'pname3'
,pvalue3)
```

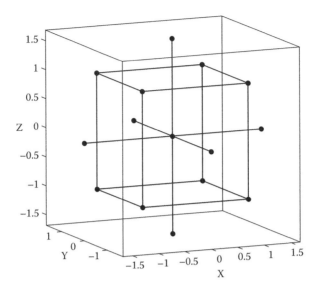

FIGURE 8.2   Three-factor circumscribed central composite design.

where **'pname1'**, can be either **'center'**, the number of center points, or **'uniform'** to select the number of center points to give uniform precision, or **'orthogonal'** (the default) to give an orthogonal design. **pvalue1** is the number of points. **'pname2'** is **'fraction'**. **pvalue2** is the fraction of the full factorial for the cube portion, expressed as an exponent of 1/2 (0 = whole design (the default), 1 = ½ fraction, 2 = 1/4 fraction, etc.). **'pname3'** is **'type'**. **pvalue3** is either **'inscribed'**, **'circumscribed'**, or **'faced'**.

For example, the statement

```
dCC2 = ccdesign(3,'center',3,'type','faced')
```

typed into the Command Window generates the following design.

```
dCC2 =

    -1    -1    -1
    -1    -1     1
    -1     1    -1
    -1     1     1
     1    -1    -1
     1    -1     1
     1     1    -1
     1     1     1
    -1     0     0
     1     0     0
     0    -1     0
     0     1     0
     0     0    -1
     0     0     1
     0     0     0
     0     0     0
     0     0     0
```

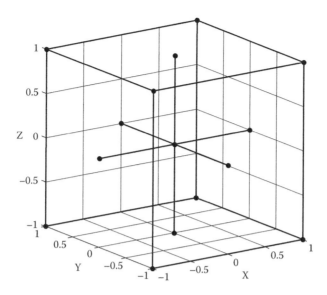

FIGURE 8.3 Three-factor faced central composite design.

This full factorial design has 3 factors, 3 center points, and is the faced type. The design can be visualized in Figure 8.3.

8.2.2 BOX-BEHNKEN DESIGNS

The Box-Behnken designs differ from central composite designs in that the parameter values are at the midpoints of edges of the design space and at the center. There are no corner points and all factors are scaled between −1 and 1 (Figure 8.4).

The MATLAB statement **dBB = bbdesign(n)** generates a Box-Behnken design for **n** factors; **n** must be an integer **3** or larger. The output matrix **dBB** is m-by-**n**, where m is the number of runs in the design. Each row represents one run, with settings for all factors represented in the columns. Factor values are normalized so that the cube points take values between −1 and 1.

One also can specify the number of center points for a Box-Behnken design as with central composite designs. The MATLAB statement has the following syntax

dBB=bbdesign(nfactors,'center',pvalue1)

where **pvalue1** is the number of center points. For example, the Box-Behnken design with three factors and three center points is generated by typing this statement in the Command Window:

dBB=bbdesign(3,'center',3).

The resulting output in the Command Window is:

dBB =

```
        -1      -1       0
        -1       1       0
         1      -1       0
         1       1       0
```

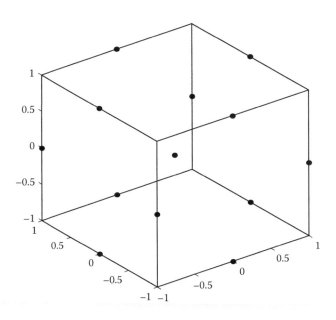

FIGURE 8.4 Example of a three-factor Box-Behnken experimental design.

-1	0	-1
-1	0	1
1	0	-1
1	0	1
0	-1	-1
0	-1	1
0	1	-1
0	1	1
0	0	0
0	0	0
0	0	0

Example 8.1

Table 8.2 shows the parameter values for the striped bass egg mortality data (Hall et al. 1981) and their normalized values for each run. For ΔT, the normalized value is $x_1 = (\Delta T - 6) / 4$, for total residual chlorine (TRC) it is $x_2 = (TRC - 0.15) / 0.15$, and for exposure time (EXP) it is $x_3 = (EXP - 2) / 2$.

Note that the center point (0,0,0) is replicated three times to improve the precision of the experiment. This design is a modified Box-Behnken design, which contains some corner points to improve precision of the response surface in those areas of the design space (Figure 8.5).

This design is used in a response surface model described in Chapter 9.

TABLE 8.2
Factor Values and Their Normalized Values for Striped Bass Egg Mortality Experiment

Run	ΔT	TRC	EXP	x_1	x_2	x_3
1	2	0.00	2.0	−1	−1	0
2	6	0.00	0.08	0	−1	−1
3	6	0.00	2.0	0	−1	0
4	10	0.00	4.0	+1	−1	+1
5	2	0.15	2.0	−1	0	0
6	2	0.15	4.0	−1	0	+1
7	6	0.15	0.08	0	0	−1
8	6	0.15	2.0	0	0	0
9	6	0.15	2.0	0	0	0
10	6	0.15	2.0	0	0	0
11	6	0.15	4.0	0	0	+1
12	10	0.15	0.08	+1	0	−1
13	10	0.15	2.0	+1	0	0
14	2	0.30	0.08	−1	+1	−1
15	6	0.30	2.0	0	+1	0
16	6	0.30	4.0	0	+1	+1
17	10	0.30	2.0	+1	+1	0

Source: From L. W. Hall Jr., D. T. Burton, S. L. Margrey, and K. R. Dixon. 1981. "Time-Related Mortality Responses of Striped Bass (*Morone saxatilis*) Ichthyoplankton after Exposure to Simulated Power Plant Chlorination Conditions." *Water Research* 15:903–910. With permission from Elsevier.

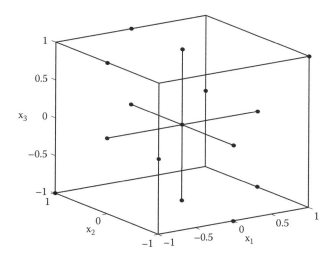

FIGURE 8.5 Experimental design of L. W. Hall Jr., D. T. Burton, S. L. Margrey, and K. R. Dixon. 1981. "Time-Related Mortality Responses of Striped Bass (*Morone saxatilis*) Ichthyoplankton after Exposure to Simulated Power Plant Chlorination Conditions." *Water Research* 15:903–910 with combined central composite and Box-Behnken design elements.

TABLE 8.3
Experimental Conditions for Each of the Treatments

Cube Points				Center Points				Star Points			
Code	pH	H	DOC	Code	pH	H	DOC	Code	pH	H	DOC
K1	8	370	9.7	C1	7.25	240	21	S1	6	240	21
K2	6.5	370	9.7	C2	7.25	240	21	S2	7.25	35	21
K3	8	110	32.3	C3	7.25	240	21	S3	7.25	240	40
K4	6.5	110	32.3					S4	7.25	240	2
K5	8	110	9.7					S5	7.25	445	21
K6	6.5	110	9.7					S6	8.5	240	21
K7	6.5	370	32.3								
K8	8	370	32.3								

Source: Data from D. G. Heijerick, C. R. Janssen, and W. M. De Coen. 2003. "The Combined Effects of Hardness, pH, and Dissolved Organic Carbon on the Chronic Toxicity of Zn to *D. magna*: Development of a Surface Response Model." *Archives of Environmental Contamination and Toxicology* 44:210–217. With permission from Springer.

Note: Hardness is expressed as mg $CaCO_3$/L; DOC as mg/L.

EXERCISES

The data in Table 8.3 show the experimental conditions for a series of experiments on the combined effects of hardness (H), pH, and dissolved organic carbon (DOC) on the chronic toxicity of Zn to *D. magna* (from Heijerick et al. 2003).

1. Is this a central composite design or a Box-Behnken design?
2. Create a new table with the added normalized values of pH, hardness, and DOC.
3. Using the MATLAB function **ccdesign**, generate the design for this experiment.

REFERENCES

Box, G. E. P., W. G. Hunter, and J. S. Hunter. 1978. *Statistics for Experimenters: An Introduction to Design, Data Analysis, and Model Building.* New York: Wiley.

Box, G. E. P., J. S. Hunter, and W. G. Hunter. 2005. *Statistics for Experimenters: Design, Innovation, and Discovery, 2nd Edition.* Hoboken, NJ: Wiley.

Hall, L. W., Jr., D. T. Burton, S. L. Margrey, and K. R. Dixon. 1981. "Time-Related Mortality Responses of Striped Bass (*Morone saxatilis*) Ichthyoplankton after Exposure to Simulated Power Plant Chlorination Conditions." *Water Research* 15:903–910.

Heijerick, D. G., C. R. Janssen, and W. M. De Coen. 2003. "The Combined Effects of Hardness, pH, and Dissolved Organic Carbon on the Chronic Toxicity of Zn to *D. magna*: Development of a Surface Response Model." *Archives of Environmental Contamination and Toxicology* 44:210–217.

Hunter, J. S., and T. H. Naylor. 1970. "Experimental Designs for Computer Simulation Experiments" *Management Science* 16:422–434.

Martin, F. F. 1968. *Computer Modeling and Simulation.* New York: Wiley.

9 Analysis of Simulation Experiments

9.1 SIMULATION OUTPUT ANALYSIS

9.1.1 TYPES OF SIMULATIONS

There are two types of simulations depending upon the purpose of the simulation. If we are interested in the dynamics of an animal population exposed to a toxicant over one year or the dynamics of a toxicant taken up by plants throughout a growing season, there is a natural terminating point to the simulation. This type of simulation is defined as a *terminating simulation*. If there is no natural ending point to the simulation, we must make a decision on when to end the *nonterminating simulation*. Determining the starting point and the termination point will affect the type of output analysis.

For example, in a nonterminating simulation we might be interested in whether the system reaches equilibrium. How long should we run the simulation to determine whether the system reaches equilibrium, or the rate at which the system approaches equilibrium? There also may be naturally occurring cycles within the simulation period such as a diel cycle within a yearly simulation.

9.1.2 OUTPUT ANALYSIS METHODS

Analysis of simulation output uses the same statistical methods as those for analyzing data from laboratory or field experiments. The most common statistical measures are the mean, variance or standard deviation, and confidence intervals for the mean. Suppose we run a series of n simulations of length m, using a stochastic model (Figure 9.1). For a state variable Y in the model, each run j at each time step i generates a random value y_{ji} for that variable. For example, using the Weiss and Kavanau (1957) growth model, we generated a series of five runs (Figure 9.2). The deterministic model used in Example 5.7 was made stochastic by making maximum growth rate, $\log(2)$, a random variable. We assumed a normal distribution with mean $\log(2)$ and standard deviation 0.5.

We can estimate the mean and confidence intervals for each time step or for each run. The means for each time step are straightforward because the runs are independent (assuming that a different sequence of random numbers is used for each run). However, if the simulation output exhibits initial transient behavior, the values Y_i for a single run may not be independent. The mean for a run then would violate the assumption of independence. The mean ordinarily would be useful only for the portion of the time series that shows steady-state behavior, if any, which would be made up of independent Y_i values. Therefore, we must determine the steady-state portion of the output, if it exists, and estimate the mean for that time series.

To estimate the mean at each time step, we sum over runs:

$$\bar{Y}_i = \frac{\sum_{j=1}^{n} y_{ji}}{n} \tag{9.1}$$

$$y_{11}, \cdots, y_{1i}, \cdots, y_{1m} \quad \bar{y}_1$$

$$y_{21}, \cdots, y_{2i}, \cdots, y_{2m} \quad \bar{y}_2$$

$$\vdots$$

$$y_{n1}, \cdots, y_{ni}, \cdots, y_{nm} \quad \bar{y}_n$$

$$\overline{\bar{y}_1, \cdots, \bar{y}_i, \cdots, \bar{y}_m}$$

FIGURE 9.1   Example of simulation output with n runs of length m. Run means are to the right of the vertical bar. Time-step means are below the horizontal bar.

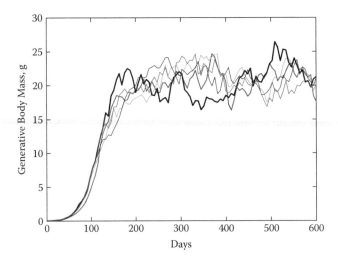

FIGURE 9.2   Generative mass output from five runs of the Weiss and Kavanau (1957) growth model with stochastic growth rate.

The standard formula for $100(1 - \alpha)$ percent confidence intervals applies:

$$\bar{X} \pm t_{n-1,1-\alpha/2} \frac{S}{\sqrt{n}} \tag{9.2}$$

where

$\bar{X}$ = the time-step mean

$t_{n-1,1-\alpha/2}$ = student's t statistic with $n - 1$ d.f. and critical value α

S = sample standard deviation

The problem of the initial transient in estimating run means involves a procedure called *warming up the model*. In this procedure, the output from several simulation runs are plotted to identify the point l where the transient behavior ends. Then only those observations in the time series beyond this point are used to estimate the mean for each run. This point can be estimated visually, although more rigorous methods are available (Welch 1983, 289).

The mean for a time series from point l to the end of the simulation m, is:

$$\bar{Y}_j = \frac{\sum_{i=l}^{m} y_{ji}}{m - l} \tag{9.3}$$

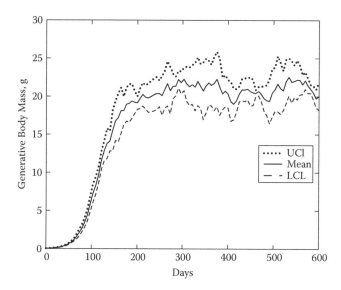

FIGURE 9.3   Means and 95% CI at each time step for generative mass from five runs of growth model.

Example 9.1

This example also uses the Weiss and Kavanau model and the simulation output shown in Figure 9.2. Using Equations (9.1) and (9.2), we generated the mean and 95% confidence intervals at each time step for the five runs in Figure 9.2 and plotted the results in Figure 9.3.

We estimated the steady-state value of generative mass by calculating the run mean of the time-step means using Equation (9.3). We selected a value for the start of steady-state dynamics l as time step 300 and simulation termination m at time step 600. Because we sampled the output at 100 evenly distributed time points, we calculated the mean from sample point 50 to sample point 100. The resulting value is printed in the Command Window:

```
ans =
  20.9693 + 0.0005i
```

Some values of the generative mass include imaginary parts and can be ignored. The m-files for this example are **growthmeans** and the function **growth**:

```
% m-file growthmeans
% program to solve growth equations, calculate means
% and 95% confidence intervals

tspan=[0 600];
GEN = [];
y1 = [];
tic
for n = 1:5;
   T0 = [0.02418 0.0 0.02418]';
   sol = ode45(@growth,tspan,T0);
   x = linspace(0, 600,100);
   y1 = deval(sol,x,1);
   GEN = [GEN;y1];
end
```

```
sqroot=sqrt(5);
tn=2.776;
meangen = mean(GEN);
stdgen = std(GEN);
term1gen = (stdgen/sqroot)*tn;
uclgen = meangen+term1gen;
lclgen = meangen-term1gen;
plot(x,GEN)
xlabel('Days')
ylabel('Generative Body Mass, g')

figure
plot(x,uclgen,x,meangen,x,lclgen)
xlabel('Days')
ylabel('Generative Body Mass, g')
legend('UCl','Mean','LCL','Location','East')
toc

% function  growth
% growth model from Weiss and Kavanau 1957
function Tdot = growth(t,T)
k1=0.5077;          %rate constant for conversion of G to D
k2=0.1154;          %rate constant for maintenance of D
k3=0.0089;          %rate constant for the catabolic loss of D
Go=0.02418;         %initial generative mass of the zygote
Ge=4096*Go;         %maximum adult generative mass
b=0.8335;           %ratio between feedback at equilibrium and
                    %complete inhibition
n=0.5;
r=log(2)+0.15*randn;
term1=1-(b*(T(1)^n-Go^n)./(Ge^n-Go^n));
Gdot=(T(1)*r)*term1-k1*T(1)*term1-k2*T(1);
Ddot=k1*T(1)*term1+k2*T(1)-k3*T(2);
Mdot=(T(1)*r)*term1-k3*T(2);
Tdot=[Gdot; Ddot; Mdot];
```

We may want to estimate a mean and CI with a predetermined precision defined by the relative error γ of the mean $\bar{X}$, where $\gamma = |\bar{X} - \mu| / |\mu|$. In other words, we would like the difference between the mean of our simulation runs and the true mean of all possible simulations relative to the population mean to be within γ, where γ is between 0 and 1. As with all experimental statistics, precision of the estimated mean (and a decreased CI) can be improved by increasing the sample size. This sequential procedure involves increasing the number of simulation runs until the mean and CI meet our relative error criterion (Law 2007):

1. Run n replications of the simulation.
2. Compute $\bar{X}$ and the confidence limit (CL) from the X_{ij} at each time step.
3. If $CL / |\bar{X}| \le \gamma / (1-\gamma)$, use the resulting $\bar{X}$ and CI, otherwise increase the number of runs and repeat the procedure.

9.2 STABILITY ANALYSIS

Stability analysis is used to analyze system behavior in response to perturbations. A stable system is one that remains in a steady state unless perturbed by external disturbances and returns to a stable state once the disturbance is removed. The external disturbances, besides being referred to as *perturbations*, also have been called *control variables, stimuli*, or *forcing functions*. A more formal definition is stated in the next section.

9.2.1 LINEAR SYSTEMS

The definition of stability for linear systems has two parts. One part refers to the response of the system to an impulse and the second refers to the response to input that remains within specified bounds.

Definition 9.1: A linear system is stable if its impulse response approaches zero as time approaches infinity or, a system is stable if every bounded input produces a bounded output (BIBO). ∎

The definition of stability of nonlinear systems differs somewhat from that for linear systems. The stability of linear systems can be analyzed by determining the system response to the input of certain *singularity* functions. Two such functions are the *unit step function* and the *unit impulse function*. The impulse response to a disturbance is the response to the impulse function, which is used to determine stability according to the first part of Definition 9.1. Before defining the unit impulse function, we need to define the other function—the unit step function:

$$u(t - t_0) = \begin{cases} 1 & \text{for} \quad t > t_0 \\ 0 & \text{for} \quad t \leq t_0 \end{cases} \tag{9.4}$$

This function remains at zero as t increases until time t_0 when the function jumps to 1 (Figure 9.4). The unit impulse function now can be defined in terms of the unit step function. This function takes the unit step function and reduces the time the function is at 1 to a small amount of time Δt:

$$\delta(t) = \lim_{\Delta t \to 0} \left[\frac{u(t) - u(t - \Delta t)}{\Delta t} \right] \tag{9.5}$$

where $u(t)$ is the unit step function. Figure 9.5 shows a graph of the unit impulse function. In practice, because $1/\Delta t$ cannot get close to infinity, the unit impulse function can be approximated by a pulse over a short time step. Both the unit step function and the unit impulse function can be used as inputs to a model, representing external perturbations. The actual value of the perturbation is the function multiplied by a constant.

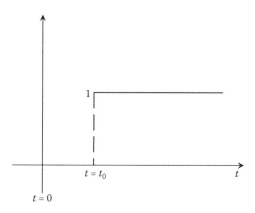

FIGURE 9.4 Unit step function. (From J. J. DiStefano, III, et al., 1990. *Feedback and Control Systems*, 2nd ed., Schaum's Outline Series, New York: McGraw-Hill. With permission.)

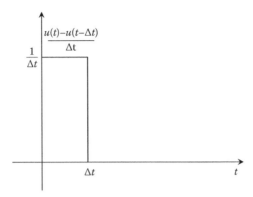

FIGURE 9.5   Unit impulse function. (From J. J. DiStefano, III, et al., 1990. *Feedback and Control Systems*, 2nd ed., Schaum's Outline Series, New York: McGraw-Hill. With permission.)

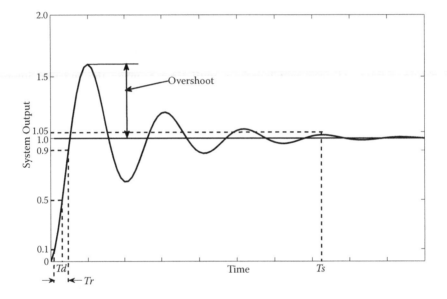

FIGURE 9.6   Example of a unit step response. (From J. J. DiStefano, III, et al., 1990. *Feedback and Control Systems*, 2nd ed., Schaum's Outline Series, New York: McGraw-Hill. With permission.)

Now we can define stability in terms of the system response to the unit impulse function (or its approximation). A formal definition of the impulse response is:

Definition 9.2: The impulse response of a linear system is the output of the system when the input is an impulse function and all initial conditions are zero. ∎

The response to the singularity functions has a steady-state response and a transient response (see Section 2.2.1.2). A system response to a unit step function is shown in Figure 9.6 (DiStefano et al. 1990). The transient response is seen at the beginning of the response. The response then shows damped oscillations before converging on the steady-state solution. Figure 9.6 illustrates several measures of the magnitude and speed of the response:

1. **Overshoot:** The *overshoot* measures the maximum difference between the transient response and the steady-state response. It can be described as a percentage of the steady-state solution.
2. **Delay Time Td:** The delay time *Td,* is the time for the response to go from zero to one-half the steady-state solution.
3. **Rise Time Tr:** The rise time *Tr,* is the time it takes for the response to go from 0.1 to 0.9 times the steady-state solution.
4. **Settling Time Ts:** The settling time *Ts,* is the time it takes the response to reach a specified percentage of the steady-state solution, such as 5%.

9.2.2 Nonlinear Systems

The response analysis described previously has not been developed exactly for nonlinear systems, although the responses to singularity functions still can be informative. Because the response of nonlinear systems can differ significantly from that of linear systems, we must use a different set of stability criteria. One definition of stability of nonlinear systems, the Lyapunov stability criterion, developed by the Russian mathematician Aleksandr Lyapunov, assumes that the system can be described by a set of first-order nonlinear difference or differential equations, the solution has equilibrium, and there is no external forcing function.

Definition 9.3: A system is stable if its output begins within a specified distance of its equilibrium and never exceeds that distance, and a system is asymptotically stable if system output approaches its equilibrium as time approaches infinity. ∎

9.2.3 Relative Stability

Once we have determined that a system is stable, the next question is how stable is it? Is it likely to remain stable if perturbed or is it more likely to become unstable, either crashing or expanding indefinitely? This is the question of relative stability. Most of the work on relative stability has been in control systems, particularly in signal processing where stability is measured as a function of the response of a control system to inputs with varying frequencies. There has been little application of frequency response measures of relative stability in natural resource systems. One example of ecosystem frequency response analysis, however, is based on a model of calcium cycling in a tulip poplar (*Liriodendron tulipifera*) forest in Tennessee (Shugart et al. 1976).

The objective of a relative stability analysis is to determine the range of perturbations to the system over which the system remains stable. One such measure of relative stability was developed by Patten and Witkamp (1967) to compare the stability of different compartments in an ecosystem (see Example 6.5). This metric, σ, could be used, however, to assess stability relative to the point at which a system becomes unstable:

$$\sigma = \frac{x_{j(eq)} \cdot \delta_j t}{\Delta x_j \cdot \Delta_j t} \tag{9.6}$$

where
$x_{j(eq)}$ = the equilibrium concentration (mass) of compartment or subsystem j
$\delta_j t$ = the duration of the perturbation
Δx_j = the perturbation in concentration (mass) of compartment or subsystem j
$\Delta_j t$ = settling time, T_s

This system statistic measures stability such that σ decreases as the factors in the denominator increase. Therefore, stability decreases as the magnitude of the perturbation increases. The system also will show increased instability with an increase in the time to return to equilibrium (settling time) relative to the duration of the disturbance.

9.2.4 RESILIENCE

Related to the concept of system stability is system resilience (Holling 1973). Resilience has more to do with the structural integrity of the system than whether or not it is stable. This measure of persistence is particularly relevant to systems perturbed by toxic substances. It depends on whether the system is able to maintain relationships among populations or state variables following perturbations from driving or controlling variables. According to Holling, the balance between resilience and stability depends upon the history of the system as it evolved in adapting to the range of disturbances experienced. As such, systems experiencing a wide range of perturbations may show low stability but high resilience. Measures of resilience should reflect the structural components of the system and their relationships. For example, how much of an increase or decrease in parameter values is necessary before one or more state variables goes to extinction? This could be considered a boundary or special case test of model verification (Section 2.3.2.3) and an example of a simulation to identify unanticipated effects (Section 1.2.9).

9.3 SENSITIVITY ANALYSIS

A measure of model sensitivity is the amount that the system output differs from its nominal value when one of its parameters differs from its nominal value. In mathematical terms, sensitivity can be expressed as a partial derivative of the state variable x with respect to a model parameter p:

$$S = \frac{\partial x}{\partial p} \tag{9.7}$$

As a discrete approximation, the following equation is a dynamic version of one described by Haefner (2005):

$$S(t) = \frac{\dfrac{(X_a(t) - X_n(t))}{X_n(t)}}{\dfrac{(P_a(t) - P_n(t))}{P_n(t)}} \tag{9.8}$$

where
   $S(t)$ = the parameter sensitivity at time t
   $X_a(t)$ = the state variable with the adjusted parameter value
   $X_n(t)$ = the state variable with the normative parameter value
   $P_a(t)$ = the adjusted parameter value
   $P_n(t)$ = the normative parameter value

It is important to remember that a model that is found to be sensitive to a parameter does not necessarily mean that the real-world system will be as sensitive to the comparable real-world parameter. It should be used only as a guide for investigating that parameter in the real-world system. Of course finding a sensitive model parameter that does not have a real-world counterpart is not very useful. This is one reason that models should be as realistic as possible, considering the purpose of the model.

Example 9.2

In this example, we used the Weiss and Kavanau (1957) model to estimate the sensitivity of the state variable generative body mass, G, to two parameters: the ratio between feedback at equilibrium and complete inhibition, b, and the maximum adult G mass, G_e. We wrote two functions to model the body mass, one for the normative variables and parameters, **growthn**, and one for the adjusted variables and parameters, **growtha**, which are similar to the function **growth** in Example 9.1. We then wrote two m-files to solve the differential equations, again, one for the normative variables and parameters, **growth2n**, and one for the adjusted variables and parameters, **growth2a**, which are similar to the m-file, **growthmeans**, in Example 9.1. In these m-files, we create output of the same length, so that we can calculate the differences between normative and adjusted variables as in Equation 9.8. In each of these m-files, we save the variable G to a **.mat** file using the **save** command. The last step in estimating sensitivity is to write an m-file, **sensitivity**, to calculate S. This requires loading the **.mat** files using the **load** command. The estimated sensitivities for the parameters G_e and b are shown in Figure 9.7.

The m-file programs for the adjusted variables and parameters, as well as for sensitivity, are listed below.

```
% function growtha for adjusted variables and parameters
% growth model from Weiss and Kavanau 1957
function Tadot = growtha(~,Ta)
k1=0.5077; k2=0.1154; k3=0.0089;
Go=0.02418;              %initial generative mass of the zygote
Ge=4096*Go; De=53190*Go;
b=0.8335;
n=0.5;
Gea=1.1*Ge;
ba=1.1*b;

%term1=1-(b*(Ta(1)^n-Go^n)./(Gea^n-Go^n));
term1=1-(ba*(Ta(1)^n-Go^n)./(Ge^n-Go^n));
Gadot=(Ta(1)*log(2))*term1-k1*Ta(1)*term1-k2*Ta(1);
Dadot=k1*Ta(1)*term1+k2*Ta(1)-k3*Ta(2);
Madot=(Ta(1)*log(2))*term1-k3*Ta(2);
Tadot=[Gadot; Dadot; Madot];
```

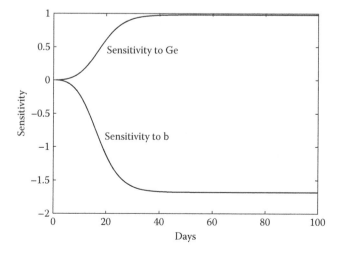

FIGURE 9.7 Sensitivity of generative body mass to parameters b and Ge.

```
% m-file growth2a
% program to solve adjusted growth equations
T0=[0.02418 0.0 0.02418]';
tspan=[0 850];
y1a = [];
y2a = [];

    sol = ode45(@growtha,tspan,T0);
    x = linspace(0, 600,100);
    y1a = deval(sol,x,1);
    y2a = deval(sol,x,2);

plot(x,y1a)
hold on
plot(x,y2a)

legend('Generative Mass','Differentiated Mass',...
        'Total Mass','Location','East')
xlabel('Days')
ylabel('Body Weight, g')
% Save all variables from the workspace to testa.mat:
save C:\MATLAB\testa y1a

%m-file sensitivity
%program to estimate sensitivity to model parameters
load C:\MATLAB\testa y1a
load C:\MATLAB\testn y1n

Gn = y1n;
Ga = y1a;
k1=0.5077; k2=0.1154; k3=0.0089;
Go=0.02418;           %initial generative mass of the zygote
Ge=4096*Go; De=53190*Go;
b=0.8335;
n=0.5;

Gea = 1.1*Ge;
ba = 1.1*b;
%S = ((Ga-Gn)./Gn)./((Gea-Ge)./Ge); %sensitivity to Ge
S = ((Ga-Gn)./Gn)./((ba-b)./b);       %sensitivity to b

plot(S)
xlabel('Days')
ylabel('Sensitivity')
hold on
```

9.4   RESPONSE SURFACE METHODOLOGY

Up to this point, we have considered models and simulation as end points themselves. We develop a model, implement it on a computer, and run simulation experiments to predict the response of a system to perturbations. One form of analysis of simulation experiments that can simplify the prediction process, evaluate parameter influence, and identify maximum (or minimum) responses as a function of select parameters is *response surface methodology* (RSM). The method was introduced in 1951 by the British statistician, George E. P. Box, and the British chemist, K. B. Wilson, while working at Imperial Chemical Industries (ICI) in Manchester, England (Box and Wilson 1951).

The first step in the procedure is to run a series of simulations using a factorial design or fractional factorial design, to identify the most significant explanatory variables. Once the significant explanatory variables have been identified, a more focused design, such as a central composite design (Section 8.2.1), is used to determine the responses to the explanatory variables within a particular range of interest. Second, using least squares regression to fit a polynomial, we generate the response-surface model. This model then can be used to predict responses for values of the explanatory variable and parameters that were not simulated using the original model, including an optimum (maximum or minimum) value. The response-surface model is only an approximation because it does not contain all the information of the original model; however, the RSM model is easier to apply.

For example, the second-degree polynomial for three explanatory variables x_1, x_2, and x_3, has the form

$$\hat{y} = \beta_0 + \beta_1 x_1 + \beta_2 x_2 + \beta_3 x_3 + \beta_{12} x_1 x_2 + \beta_{13} x_1 x_3 + \beta_{23} x_2 x_3 +$$
$$\beta_{11} x_1^2 + \beta_{22} x_2^2 + \beta_{33} x_3^2$$

(9.9)

where the terms on the right-hand side (RHS) include, from left to right, an intercept, linear terms, interaction terms, and squared terms.

In MATLAB, a response surface model can be generated using the least squares regression procedure, **regstats** (see Section 7.1.3). If the response is quadratic, then the **'quadratic'** option is specified. Once the parameters in the model have been estimated, they are included in a model structure such as Equation (9.9). The resulting model can be plotted in three dimensions to display the response surface using the MATLAB function, **scatter3**. First, each of the explanatory variables is mapped onto one of the three axes. The volume defined by the three axes then is filled by the values of the response variable.

Following our previous example, where we used three explanatory variables to generate a three-dimensional volume, we then can explore the response to just two of the variables by holding the third variable constant. Then we can plot the *two-dimensional response*. This procedure is simple and straightforward using the MATLAB function **slice**:

```
slice(X,Y,Z,V,Sx,Sy,Sz,'method')
```

where **X,Y,Z,** are the three axes defined for the response surface model, **V** is the three-dimensional volume defined by the model, and **Sx**, **Sy**, and **Sz** define the points along the three axes where the slices are drawn. The color at each point in the slice is determined by 3-D interpolation into the volume **V**. Interpolation **'method'** can be **'quadratic'**, **'cubic'**, or the default, **'linear'**.

MATLAB provides an interactive graphical user interface (GUI) for fitting and visualizing a polynomial response surface called RSTOOL that is initiated by typing **rstool** in the Command Window. For a specific model, we need to specify variables, their names, and the type of model being fitted to the data. The MATLAB command **rstool(x,y,model,alpha,xname,yname)** opens the GUI and specifies the predictor variables in **x**, the response variable **y**, and the type of regression **model**, i.e., whether it is **'linear'** (constant and linear terms [the default]), **'interaction'** (constant, linear, and interaction terms), **'quadratic'** (constant, linear, interaction, and squared terms), or **'purequadratic'** (constant, linear, and squared terms). Including the option **alpha** plots 100(1-**alpha**)% confidence intervals for predictions. **xname** and **yname** labels the axes using the names in the strings **xname** and **yname**. RSTOOL displays a family of plots in green, one for each combination of columns in **x** and **y**. RSTOOL plots a 95% global confidence interval for predictions as two red curves. You can drag the dashed blue reference line to examine predicted values.

Example 9.3

This example uses the modified Box-Behnken design described in Example 8.1, Table 8.2 (Hall et al. 1981). We used the following MATLAB statement to fit a quadratic model to the data where mortality (**mort**) is the response variable and total residual chlorine (TRC), temperature increase (deltaT), and exposure time (EXP) are the predictor variables found in the input matrix **basseggdata**.

```
stats = regstats(mort,basseggdata,'quadratic','beta')
```

The model was specified with the following statement:

```
morthat = b(1) + b(2)*X1 + b(3)*X2 + b(4)*X3 + ...
       b(5)*X1.*X2 + b(6)*X1.*X3 + b(7)*X2.*X3 + ...
       b(8)*X1.^2 + b(9)*X2.^2 + b(10)*X3.^2;
```

using the parameters **b(1)** to **b(10)** generated from the **regstats** command and where **X1**, **X2**, and **X3** are TRC, deltaT, and EXP, respectively. We used the **scatter3** function to plot the response volume using the statement:

```
hmodel = scatter3(X1(:),X2(:),X3(:),6,morthat(:),'filled');
```

and the observed data using the following statement:

```
hdata = scatter3(TRC,deltaT,Exp,'ko','filled');
```

The results show a general increase in mortality from the point 0,0,0 to the maximum values of the predictor variables (Figure 9.8).

We explored the relationship of egg mortality to each of the response variables two at a time instead of all three, using the slice function. For example, we set deltaT (X2) near its midpoint (5) and plotted the response to TRC and EXP (Figure 9.9) using the following statements:

```
X2slice = 5;
slice(X1,X2,X3,morthat,[],X2slice,[])
```

We used the default **'linear'** option.

We used RSTOOL to examine further the relationship of egg mortality to each of the predictor variables separately. First we set the alpha level to 0.01 and then generated the GUI with the following statements:

```
alpha = 0.01; % Significance level
rstool(basseggdata,mort,'quadratic',alpha,xn,yn)
```

The resulting plots are shown in Figure 9.10. The results of both the splice and RSTOOL show a linear increase in egg mortality as a function of TRC. According to RSTOOL there also is a linear increase with deltaT. The response to EXP, however, appears to be quadratic.

The commands for all these procedures are included in the m-file **eggresponse**.

```
% program eggresponse
% m-file to estimate a response surface model
% for striped bass egg mortality as a function
% of total residual chlorine (TRC), increase in
% temperature (deltaT), and exposure time (EXP)

% load matrix of explanatory variables
[basseggdata,txt]=xlsread('stripedbasseggs2.xls','Sheet1','A2:C18');
```

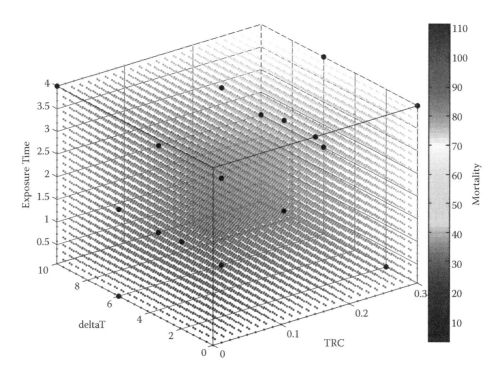

FIGURE 9.8 Response volume of striped bass egg mortality as a function of TRC, deltaT, and EXP. (Data from L. Hall, et al., 1981. "Time-Related Mortality Responses of Striped Bass (*Morone saxatilis*) Ichthyoplankton after Exposure to Simulated Power Plant Chlorination Conditions," *Water Research* 15: 903–910. With permission from Elsevier.)

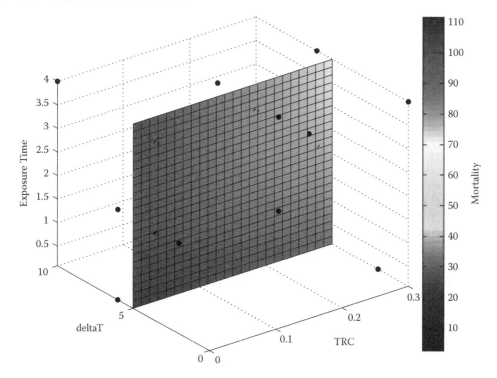

FIGURE 9.9 Plot of egg mortality as a function of TRC and EXP at the deltaT value of 5.

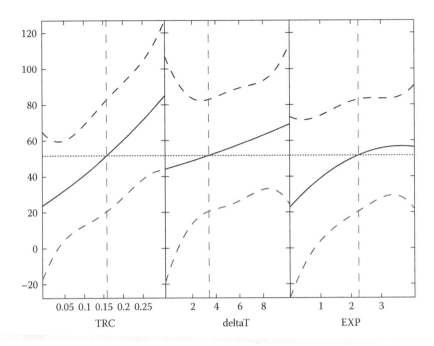

FIGURE 9.10 Plots of striped bass egg mortality as functions of TRC, deltaT, and EXP.

```
TRC = basseggdata(:,1);
deltaT = basseggdata(:,2);
Exp = basseggdata(:,3);
mort = [22.832 27.374 32.97 42.02 13.916 32.528...
        51.802 47.644 64.648 86.952 92.298 48.898...
        59.406 57.848 100.00 42.326 89.382]';

% generate the response surface model using regstats
stats = regstats(mort,basseggdata,'quadratic','beta');
b = stats.beta; % Model coefficients
xx1 = linspace(min(TRC),max(TRC),25);
xx2 = linspace(min(deltaT),max(deltaT),25);
xx3 = linspace(min(Exp),max(Exp),25);

[X1,X2,X3] = meshgrid(xx1,xx2,xx3);

morthat = b(1) + b(2)*X1 + b(3)*X2 + b(4)*X3 + ...
          b(5)*X1.*X2 + b(6)*X1.*X3 + b(7)*X2.*X3 + ...
          b(8)*X1.^2 + b(9)*X2.^2 + b(10)*X3.^2;

% plot observed values and predicted response volume
hmodel = scatter3(X1(:),X2(:),X3(:),6,morthat(:),'filled');
hold on
hdata = scatter3(TRC,deltaT,Exp,'ko','filled');
axis tight
xn = {'TRC' 'deltaT' 'Exp'}
yn = ['mortality']
xlabel('TRC')
ylabel('Delta T')
zlabel('Exposure Time')
hbar = colorbar;
ylabel(hbar,'Mortality');
```

```
%delete(hmodel)                          %(remove % comment)
%X2slice = 5; % Fix deltaT               %(to run slice)
%slice(X1,X2,X3,morthat,[],X2slice,[])   %(function)


alpha = 0.01; % Significance level
rstool(basseggdata,mort,'quadratic',alpha,xn,yn)
```

EXERCISES

The data in Table 9.1 give EC_{50} values for *Daphnia magna* exposed to zinc for 21 days. Values are expressed in µg Zn/L. Treatment codes reflect the experimental conditions summarized in Table 8.3 (From Heijerick et al. 2003).

1. Create an Excel file with the data in Table 8.3. Hint: start with the m-file **eggresponse**.
2. Write an m-file to do the following:
   a. Read the Excel file.
   b. Define the variables pH, H, and DOC.
   c. Create a vector of EC_{50} values from Table 9.1.
   d. Using the **regstats** function, fit a pure quadratic model with EC_{50} as the response variable and pH, H, and DOC as the predictor variables.
   e. Output the statistics beta (the model parameter values), yhat (the predicted EC_{50} values), Student's *t* values, standard errors, and *p* values for each parameter estimate, the r-square and adjusted r-square values.
   f. Plot the predicted EC_{50} values over the observed values. Include a line with slope = 1.
   g. Define the 3-D volume for the RSM model using the **meshgrid** function.
   h. Define the RSM model using the parameter values from the **regstats** function output and the **meshgrid** volume.

TABLE 9.1
EC_{50} Values (95% Confidence Limits) for *Daphnia magna* Exposed to Zinc for 21 Days

EC_{50} (95% C.L.)	EC_{50} (95% C.L.)
269 (226–325)	644 (391–1544)
207 (189–238)	610 (564–660)
823 (756–894)	465 (438–603)
476 (442–515)	413 (382–447)
346 (304–402)	1019 (997–1043)
350 (335–365)	245 (231–285)
688 (617–759)	520 (438–595)
828 (787–867)	991 (907–1077)
555 (503–615)	

Source: Data from D. Heijerick et al., 2003. "The Combined Effects of Hardness, pH, and Dissolved Organic Carbon on the Chronic Toxicity of Zn to *D. magna*: Development of a Surface Response Model." *Archives of Environmental Contamination Toxicology* 44:210–217. With permission from Springer.

Note: Values are expressed in µg Zn/L.

---

    i. Create the 3-D volume of the predicted and observed EC_{50} values using the **scatter3** function. Include a color bar for the EC_{50} values.

    j. Create slices of the volume at pH = 7, H = 200, and DOC = 20, using the **slice** function.

3. Open the **rstool** in the Command Window.

    a. Observe the changes in the EC_{50} values over the range of hardness values when (1) DOC = 10mg/L and pH = 6.5, and (2) DOC = 35 mg/L and pH = 8.

    b. Observe the changes in the EC_{50} values over the range of DOC values when (1) hardness = 100 mg/L and pH = 6.5, and (2) hardness = 400 mg/L and pH = 8.

    c. Observe the changes in the EC_{50} values over the range of pH values when (1) hardness = 100 mg/L and DOC = 10 mg/L, and (2) hardness = 400 mg/L and DOC = 35 mg/L.

REFERENCES

Box, G. E. P., and K. B. Wilson. 1951. "On the Experimental Attainment of Optimum Conditions." *Journal of the Royal Statistical Society, Series B.* 13:1–47.

DiStefano, J. J., III, A. R. Stubberud, and I. J. Williams. 1990. *Theory and Problems of Feedback and Control Systems,* 2nd ed. Schaum's Outline Series, New York: McGraw-Hill.

Haefner, J. W. 2005. *Modeling Biological Systems: Principles and Applications,* 2nd ed. New York: Springer.

Hall, L. W. Jr., D. T. Burton, S. L. Margrey, et al. 1981. "Time-Related Mortality Responses of Striped Bass (*Morone saxatilis*) Ichthyoplankton after Exposure to Simulated Power Plant Chlorination Conditions." *Water Research* 15:903–910.

Heijerick, D. G., C. R. Janssen, and W. M. De Coen. 2003. "The Combined Effects of Hardness, pH, and Dissolved Organic Carbon on the Chronic Toxicity of Zn to *D. magna*: Development of a Surface Response Model." *Archives of Environmental Contamination and Toxicology* 44:210–217.

Holling, C. S. 1973. "Resilience and Stability of Ecological Systems." *Annual Review of Ecology and Systematics* 4:1–23.

Law, A. M. 2007. *Simulation Modeling and Analysis,* 4th ed. New York: McGraw-Hill.

Patten, B. C., and M. Witkamp. 1967. "Systems Analysis of [134]Cesium Kinetics." *Ecology* 48:813–824.

Shugart, H. H., Jr., D. E. Reichle, N. T. Edwards, and J. R. Kercher. 1976. "A Model of Calcium-Cycling in an East Tennessee *Liriodendron* Forest: Model Structure, Parameters, and Frequency Response Analysis." *Ecology* 57:99–109.

Weiss, P., and J. L. Kavanau. 1957. "A Model of Growth and Growth Control in Mathematical Terms." *Journal of General Physiology* 41:1–47.

Welch, P. D. 1983. "The Statistical Analysis of Computer Simulation Results." In *The Computer Performance Modeling Handbook*, ed. S. S. Lavenberg. New York: Academic Press.

10 Model Validation

As we pointed out in Chapter 2, there is considerable discussion about what constitutes a valid model and, in fact, what the definition of model validation is. A consensus definition of model validation appears below.

Definition 10.1: *Model validation* is the process of determining whether a model is an accurate representation of the system being modeled. ∎

The process of model validation involves conducting simulation experiments in which simulation output is compared with historical data collected on the real system. If the simulated output is close to the historical data, the model is considered a valid representation of the real system.

10.1 VALIDATION AND REASONS FOR MODELING AND SIMULATION

The accuracy of the model in mimicking the real system that we are willing to accept will depend upon the reason for modeling and simulation (Van Horn 1971). Perhaps the models that require the greatest validation efforts are those used for decision making and decision aiding in such areas as environmental management, hazardous waste site remediation, and risk assessment. This level of validation usually will require the greatest amount of data for model development and validation. There are numerous cases that illustrate the problems that can result with an overreliance on faulty models (Section 1.4). Similar validation efforts should be applied to modeling unanticipated effects and hypothesis and theory construction.

Perhaps a lower level of validation would be acceptable for evaluating alternative policy decisions and comparing relative predictive ability of a set of models. Models used in instruction may be designed to illustrate theories or concepts rather than make predictions, and therefore will need less rigorous validation. Since there is no real-world system to compare with a model designed for exploring nonexistent universes, the only validation possible would be the new model methods described in the following text. A similar argument can be made for system identification models.

If we are developing a new model, or we are modeling a system in which certain data cannot be collected, the model will not have data available for testing. There are, however, certain tests that can be conducted on first-time models (Hermann 1967):

Internal validity involves checking a stochastic model for low variance of outputs. Low variability is required so that variance resulting from random variable generation does not obscure changes in output resulting from changes in controlled or environmental variables.

Face validity is an evaluation of the realism of the simulation output by people knowledgeable about the real system being simulated. One type of face validity test is a *Turing* type test developed by Alan Turing, one of the inventors of the computer. In this test, output from the model and data from the real system are given to a panel of experts. If they are not able to tell which data are simulated and which are from the real system, the model passes the validity test. Sometimes this type of validity test is called a *test of the reasonableness or credibility* of the model.

Variable-parameter validity is a comparison of model parameters and variables with their real-world counterparts. It is important, therefore, to structure a model in a way such that the variables and parameters have real-world counterparts that can be measured experimentally. Another part of this validity test is sensitivity testing in which a parameter or variable is changed to determine its effect on the output (see Section 9.3).

Hypothesis validity is a comparison of the hypothesized relationships among variables and parameters in the model with their real-world counterparts. This test extends the variable-parameter validity test by examining correlations and functional relationships, not just whether real-world parameters and variables are included in the model. These relationships may be programmed in the model or just hypothesized and not explicitly programmed into the model.

Event validity is a test of the ability of a model to predict actual events, such as the mortality in a population following exposure to some toxicant. In this test, it is not necessary to actually observe such an event, only that the model is capable of predicting it. The model should be able to predict such event characteristics as magnitude and timing of the effect.

The terms *validation* and *verification* often are used interchangeably. In this text, we reserve the term validation for a test of how well the model predicts the future and whether it has made accurate predictions in the past. Verification refers to how accurately the model is represented by the model structure, either equations or block diagrams, and the computer code. If a model has several state variables, but only accurately predicts some of the variables, the model could be considered partially validated or *useful* as long as it is used only to make predictions of those variables. The relative usefulness of a model can be quantified further as model adequacy, a, and model reliability, r (Mankin et al. 1975). Suppose we have collected n_s observations from a real system and n_m comparable model predictions. Suppose further that there are n_q agreements between the real system observations and model predictions. *Adequacy* refers to the number of correct predictions relative to the total number of real-world observations:

$$a = \frac{n_q}{n_s} \tag{10.1}$$

Reliability refers to the number of correct predictions relative to the total number of model predictions:

$$r = \frac{n_q}{n_m} \tag{10.2}$$

These metrics can be used to select between two models.

10.2 TESTING HYPOTHESES

Once we have measurements from a real-world system and predictions from a model of that system, we can compare the two sets of data to see if the model appears to be doing a reasonably good job of representing such a real system. This process is generally defined as *statistical inference*. In other words, based upon our sample data from the real system and the model, can we infer that the model is "valid," that is, can we reliably use the model to make predictions about other systems or other perturbations on our real-world system? This leads us to hypothesize that the model is either valid or invalid. Consider the following two-choice hypotheses:

H_o: the null hypothesis that there is no difference between measurements taken from the real-world system and the comparable predicted values

H_a: the alternative hypothesis that there are significant differences between observed and predicted values

TABLE 10.1
Decisions about a Test of a Model's Validity and Their Consequences When the Null Hypothesis Assumes a Model Is Valid

Truth	Decision	
	Accept H_o	Reject H_o
H_o: No difference between model and real system; assume model is "valid."	*Correct* Model will produce "valid" predictions.	*Type I error* Reject valid model. Fail to make "valid" predictions.
H_a: model is significantly different from real system; model is flawed.	*Type II error* Model will produce faulty predictions.	*Correct* Model should not be used to make predictions. Revise model.

There are four possible outcomes of a test of the hypotheses. One could conclude, either correctly or incorrectly, that the null hypothesis, H_o, is true; also, one could conclude, correctly or incorrectly, that the null hypothesis is false (Table 10.1).

10.2.1 ACCEPT THE NULL HYPOTHESIS WHEN IT IS TRUE

If we accept the null hypothesis and it is true, this is the best outcome as we assume the model is valid when we found no significant difference between the real system and the model. It also illustrates the difficulty in this approach to hypothesis testing. In reality, there could be significant differences between the real system and the model, but they did not appear in our simulation results. Additional simulations could show significant differences. Our real decision is that we failed to reject a possibly faulty model.

10.2.2 REJECT THE NULL HYPOTHESIS WHEN IT IS TRUE

If we decide that the model is faulty because we failed to find no significant difference between the model and the real system, we end up rejecting the model when we could be using it to make valid predictions. This is a Type I error that occurs with probability α. The value of α is set by the experimenter, usually at 0.05 or 0.01. It should be noted that decreasing α will reduce the probability of a Type I error, but it also will result in an increase in the probability of a Type II error. If the more serious error is a Type II error, the experimenter might decide to increase the value of α to lower the probability of a Type II error.

10.2.3 ACCEPT THE NULL HYPOTHESIS WHEN IT IS FALSE

This decision assumes that the model is valid when, in fact, it is flawed. This is the most critical error we can make in drawing conclusions from the test of our model. The error is referred to as a Type II error and occurs with probability β. Unlike a Type I error, we cannot set the value of β. Its value will depend upon the real difference between the model and the real system and our choice of α (a larger value of α will lower β). It also depends upon the sample size, that is, the number of paired measurements of the real system and the model. As a result, the probability of making this type of error can be reduced by increasing the number of real system measurements and simulations.

10.2.4 REJECT THE NULL HYPOTHESIS WHEN IT IS FALSE

This is a correct decision and we conclude that the model will not produce valid predictions. Although rejecting the null hypothesis does not mean automatic acceptance of the alternative hypothesis, at

TABLE 10.2
Decisions about a Test of a Model's Validity and Their Consequences
When the Null Hypothesis Assumes a Model Is Invalid

	Decision	
Truth	Accept H$_o$	Reject H$_o$
H$_o$: Differences between model and real system are too large. Assume model is "invalid."	*Correct* Continue model development.	*Type I error* Accept bad model; make poor decisions.
H$_a$: Model has acceptable precision and accuracy. Accept model as "valid."	*Type II error* Wasted effort fixing valid model.	*Correct* Use "valid" model to make predictions.

this point, we can either discard the model or we can try to improve the model. This process would include reexamining model assumptions, relationships among variables, and parameter values.

The construct described results in a "valid" model if we accept H$_o$ when it is true. For models that require a high level of validation, such as those used to make management or policy decisions, it would be a more powerful test if we posed the null hypothesis in such a way that its rejection invalidates the model. This way of formulating null hypotheses so their rejection results in a conclusive decision is known as *strong inference* (Platt 1964). In a modeling context, the hypotheses now would be stated as:

H$_o$: there are real differences between measurements taken from the real-world system and the comparable predicted values
H$_a$: there are no significant differences between observed and predicted values

Again, the four possible outcomes of a test of the hypotheses lead to the same set of conclusions. One could conclude, either correctly or incorrectly, that the null hypothesis, H$_o$, is true; and one could conclude, correctly or incorrectly, that the null hypothesis is false (Table 10.2).

10.2.5   ACCEPT THE NULL HYPOTHESIS WHEN IT IS TRUE

This null hypothesis requires us to specify the acceptable range of differences we are willing to accept, such as a factor of two. We are making the correct decision when we accept the null hypothesis that the model is faulty and the differences between the real system and the model exceed the specified range. The resulting conclusion would be that the model needs improving as in Section 10.2.4.

10.2.6   REJECT THE NULL HYPOTHESIS WHEN IT IS TRUE

Now we have a test of model validity with *strong inference*. Rejecting the null hypothesis when it is true means we conclude the model is valid when it is not. This Type I error is the most serious because using a faulty model to make predictions could have serious consequences. However, now we can assign a value to α to minimize the probability of making a Type I error.

10.2.7   ACCEPT THE NULL HYPOTHESIS WHEN IT IS FALSE

Accepting the null hypothesis when it is false means we believe the model is faulty when it is valid. This still results in a Type II error, but now it is the lesser of the two incorrect decisions. The worst that can happen now is that we waste time trying to fix a model that is not broken.

10.2.8 Reject the Null Hypothesis When It Is False

This is the correct decision, with the desired conclusion, that we have a valid model that can be used to make accurate predictions. Again, rejecting the null hypothesis does not necessarily mean automatic acceptance of the alternative hypothesis. If we decide to accept the alternative hypothesis when it is true, we would like to maximize this probability. Rejecting H_a when it is true has a probability of β, which means accepting H_a when it is true will have a probability of $1-\beta$, called the *power* of the test. We can maximize the power of the test by minimizing β—either by reducing α or increasing sample size. Because we do not want to reduce α to avoid a Type I error, we should concentrate on increasing sample size. Since simulation experiments are relatively simple to run, the constraint on validation usually is an insufficient number of measurements from the real system.

In the next section, we present a number of statistical tests that can be used following the hypothesis construct in Table 10.1. To conduct tests under the hypothesis construct in Table 10.2, the reader is referred to Burns (2001).

10.3 STATISTICAL TECHNIQUES

Statistical techniques for testing the *goodness of fit* of simulation models have been described by Naylor and Finger (1967) and Mihram (1971):

1. *Analysis of Variance.* Analysis of variance is used to test the hypothesis that the mean of the observed time series of data is equal to the mean of the simulated series. Assumptions that must be met for analysis of variance (ANOVA) are (1) the distributions of observed and predicted values at each point in the series must come from a normal distribution, (2) successive predictive values in the series are independent of each other, and (3) the variances of the observed data and the predicted data are equal. A Student's t-test also can be used to compare observed and predicted means with the same assumptions.
2. *Chi-Square Test.* The chi-square test used to test the hypothesis that the frequencies of data generated by the simulation model in a set of output categories is the same as the frequencies of observed data in the same categories.
3. *Factor Analysis.* Separate factor analyses are run on the observed time series and that generated by the simulation model. A test is then conducted to determine whether the factor loadings are significantly different from each other.
4. *Kolmogorov-Smirnov test.* Cumulative frequency distributions are formed from both observed and simulated time series. The test is based on the sum of positive and negative differences between the two distributions. Because the distributions are empirical and not based upon any underlying parametric distribution, no assumptions concerning the distributions are required.
5. *Nonparametric Tests.* The distribution free tests are equivalent to, though less powerful than, parametric tests of the equality of means such as ANOVA and the t-test. The nonparametric equivalent of a paired t-test is the Wilcoxon signed rank test and the Kruskal-Wallis test is the nonparametric equivalent of the parametric ANOVA.
6. *Regression Analysis.* Simple linear least squares regression is run regressing simulated time-series data against observed data at the same points. The slope of the regression is tested against a slope of 1.0. The intercept also can be tested against a value of zero.
7. *Spectral Analysis.* The simulated and observed times series are converted to frequencies. Spectral analysis tests whether the frequency distributions are the same. Other forms of time-series analysis can be used to compare simulated and observed time series by comparing estimated trends and periodicity.
8. *Theil's inequality coefficient.* This test provides an index U that measures the relative differences between simulated and observed data on a scale from 0 to 1 (Theil 1970). A value of 0 indicates zero differences or perfect predictions, whereas a value of 1 indicates poor model predictions.

$$U = \frac{\sqrt{\dfrac{1}{n}\displaystyle\sum_{i=1}^{n}(P_i - A_i)^2}}{\sqrt{\dfrac{1}{n}\displaystyle\sum_{i=1}^{n}P_i^2} + \sqrt{\dfrac{1}{n}\displaystyle\sum_{i=1}^{n}A_i^2}} \tag{10.3}$$

where

P_i = predicted values
A_i = actual observed values

10.4   SOME MATLAB METHODS

We illustrate some of the statistical tests using MATLAB commands and m-files, including t-tests and ANOVA, equivalent nonparametric tests, regression, and Theil's test.

10.4.1   Paired t-test

The paired t-test is run using the MATLAB function **ttest**.

The **[h,p,ci,stats] = ttest(x,m,alpha,tail)** syntax performs a t-test to determine if a sample from a normal distribution (in **x**) could have mean **m. alpha** is the desired significance level. The **ttest** function allows specification of one- or two-tailed tests. The **tail** option is a flag that specifies one of three alternative hypotheses:

tail =   0 specifies the alternative $x \neq m$ (default)
tail =   1 specifies the alternative $x > m$
tail = −1 specifies the alternative $x < m$

By default **m** = 0, **alpha** = 0.05, and **tail** = 0

The output is **h, p, ci**, and **stats**. If **h** = 0, the null hypothesis cannot be rejected at the significance level of alpha. If **h** = 1, we reject the null hypothesis at significance level of alpha. **p** is the p-value, or the probability of observing the given result by chance given that the null hypothesis is true. Small values of **p** cast doubt on the validity of the null hypothesis. **ci** is a confidence interval for the true mean. Its confidence level is **1-alpha. stats** is a structure with two elements named "tstat" (the value of the test statistic) and "df" (its degrees of freedom).

10.4.2   Wilcoxon Nonparametric Signed Rank Test

The signed rank test is run using the MATLAB command **signrank**:

The **[p,h,stats] = signrank(x,y,alpha)** syntax returns the significance for a test of the null hypothesis that the median difference between two matched samples, **x** and **y**, is zero. **alpha** is the desired level of significance, and must be a scalar between 0 and 1. Its default value is 0.05. **p** is the probability of observing a result equally or more extreme than the one using the data (**x** and **y**) if the null hypothesis is true. If **p** is near zero, this casts doubt on this hypothesis. **h** is the result of the hypothesis test. **h** is 0 if the medians of **x** and **y** are not significantly different, and 1 if they are significantly different. **stats** is a structure with one or two fields. The field "signedrank" contains the value of the signed rank statistic. If the sample size is large, then **p** is calculated using a normal approximation and the field "zval" contains the value of the normal (Z) statistic.

TABLE 10.3
Observed and Predicted Deer Kill

	Day									
	1	**2**	**3**	**4**	**5**	**6**	**7**	**8**	**9**	**10**
Observed Deer Kill	218	206	139	127	113	76	56	58	47	33
Predicted Deer Kill	241	196	134	137	74	78	59	55	59	51
Difference, D	−23	10	5	−10	39	−2	−3	3	−12	−18

Source: Data adapted from D. Jacobs and K. R. Dixon. 1982. "A Queuing Model of White-Tailed Deer Harvest." *Journal of Wildlife Management* 46:325–352. With permission.

Example 10.1

In this example, we test the hypothesis of no difference between the observed and predicted number of deer killed on each day of hunting season using a paired t-test and the nonparametric Wilcoxon signed-rank test. The data (Table 10.3) are from a queuing model developed by Jacobs and Dixon (1982).

We wish to test the null hypothesis that the mean difference between observed and predicted deer kill is not different from zero. The paired t-test is run using the MATLAB function **ttest**:

```
[h,p,ci,stats] = ttest(x,m,alpha,tail)
```

In our example, **x** is the difference between observed and predicted deer kill, D, **m** = 0, **alpha** is 0.05, and **tail** is 0. If **h** = 0, the null hypothesis cannot be rejected at the significance level of alpha (we would conclude that no difference between observed and predicted deer kill was detected). If **h** = 1, we reject the null hypothesis at significance level of alpha (and we conclude that there was a significant difference between observed and predicted deer kill, or the model was not valid).

The second part of the example is the Wilcoxon signed-ranked test using the MATLAB command **signrank**:

```
[p,h,stats] = signrank(x,y,alpha)
```

In our example, **x** and **y** are the observed and predicted deer kill values, respectively. To implement these two tests, we use the m-file t_test:

```
%m-file t_test
%Program to test the significance of the difference between
% two means using the student's t test
day = [1 2 3 4 5 6 7 8 9 10];
obs = [218 206 139 127 113 76 56 58 47 33];
pred = [241 196 134 137 74 78 59 55 59 51];
plot(day,pred,'ro',day,obs,'b+')
xlabel ('Day of Season')
ylabel ('Deer Kill')
legend('Predicted','Observed')

D = obs - pred
Dbar = mean(D)
variance = var(D)
stddev = std(D)
[h,p,ci,stats] = ttest(D,0,0.05,0)
[p,h,stats] = signrank(obs,pred,0.05)
```

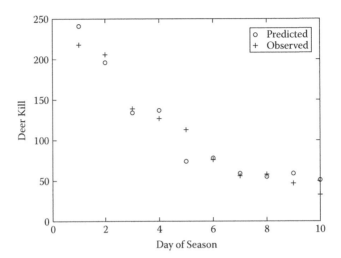

FIGURE 10.1   Observed and predicted deer kill for each day of the hunting season. (Adapted from D. Jacobs and K. R. Dixon. 1982. "A Queuing Model of White-Tailed Deer Harvest." *Journal of Wildlife Management* 46:325–352. With permission.)

The output of the m-file is Figure 10.1 and the list in the Command Window:

```
D =
   -23     10      5    -10     39     -2     -3      3    -12    -18

Dbar =
   -1.1000

variance =
  305.8778

stddev =
   17.4894

h =
    0

p =
    0.8468

ci =
  -13.6111    11.4111

stats =
    tstat: -0.1989
       df: 9
       sd: 17.4894

p =
    0.6055

h =
    0

stats =
    signedrank: 22
```

The output of the **t _ test** m-file includes the difference values, D; the mean, variance, and standard deviation of D; the output from **ttest** (**h** value; **p** value; the confidence interval for the population mean, **ci**; the t statistic; and the degrees of freedom) and the output from **signrank**. Because the t-test was not significant (**h** = 0, **p** = 0.8468, **ci** includes zero), we conclude that no difference was found between observed and predicted deer kill values. The output from **sign-rank** gives similar results. The **p** value was not significant (0.6055 > 0.05) and the conclusion is that the medians of the observed and predicted values do not differ (**h** = 0). There is no Z statistic generated because the sample size is too small.

10.4.3 LINEAR REGRESSION

One of the MATLAB functions for linear regression is **regress**. **regress** calculates multiple linear regression using least squares; however, it can be used for simple linear regression:

[b,bint,r,rint,stats] = regress(y,x,alpha) uses the input, **alpha** to calculate $100(1 - \textbf{alpha})$ confidence intervals for **b** and the residual vector, **r**, in **bint** and **rint**, respectively. The vector **stats** contains, in the following order, the R-square statistic, the F statistic and p value for the full model, and an estimate of the error variance. The F and p values are computed under the assumption that the model contains a constant term, and they are not correct for models without a constant.

Example 10.2

In this example we will use the same data in Table 10.3 to test the null hypothesis that the regression of the predicted deer kill values, **y**, on the observed deer kill values, **x**, when regressed through the origin, does not differ significantly from 1.0. To implement the **regress** command, we use the following m-file (**regress2**)

```
%m-file regress2
%Program to regress predicted values of deer kill on observed values;
obs = [218 206 139 127 113 76 56 58 47 33]';
pred = [241 196 134 137 74 78 59 55 59 51]';
plot(obs,pred,'ro')
axis([0 250 0 250])
xlabel ('Observed Deer Kill')
ylabel ('Predicted Deer Kill')
[b,bint,r,rint,stats] = regress(pred,obs,.05)

newx = 0:1:250;
yhat = b*newx;
figure
plot(obs,pred,'ro')
hold on
axis([0 250 0 250])
axis square
plot(newx,yhat)
xlabel ('Observed Deer Kill')
ylabel ('Predicted Deer Kill')
hold off
```

The output from the **regress2** m-file includes Figure 10.2 and the results from **regress**:

```
Warning: R-square and the F statistic are not well-defined unless X has
a column of ones.
Type "help regress" for more information.
> In regress at 162
  In regress2 at 8
```

```
b =
    1.0028

bint =
    0.9017    1.1039

r =
    22.3960
   -10.5708
    -5.3851
     9.6481
   -39.3131
     1.7894
     2.8448
    -3.1607
    11.8698
    17.9086

rint =
    -7.6145    52.4065
   -45.3343    24.1928
   -44.4648    33.6945
   -29.3766    48.6728
   -64.4623   -14.1638
   -39.4352    43.0140
   -38.7087    44.3984
   -44.6680    38.3466
   -28.7800    52.5196
   -21.4647    57.2819

stats =
    0.9307       NaN       NaN  307.0911
```

Note the warning about R-square and that F and p are not a number (NaN). This is a result of not estimating an intercept and forcing the regression through the origin. The resulting regression coefficient, however, is 1.0028 with an r value of 0.93 and the 95% confidence interval (CI) contains 1.0. Therefore, we can conclude that the predicted deer kill values are not significantly different from the observed values, and the model is considered valid until proven invalid. A plot of the predicted deer kill against the observed deer kill is shown in Figure 10.2.

10.4.4   Theil's Inequality Coefficient

Theil's inequality coefficient provides an index that measures the relative differences between simulated and observed data on a scale from 0 to 1. A value of 0 indicates zero differences or perfect predictions, whereas a value of 1 indicates poor model predictions. Although not a statistical test, Theil's inequality coefficient can confirm the results of statistical tests or be used where assumptions required for statistical tests cannot be met.

Example 10.3

In this example we will use the same data in Table 10.3 on the observed and predicted number of deer killed on each day of hunting season. To implement the Theil's inequality coefficient we use the Theil m-file:

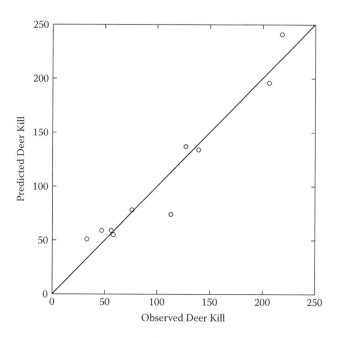

FIGURE 10.2   Observed and predicted deer kill. Diagonal line is where predicted exactly equals observed.

```
%m-file Theil
%Program to calculate Theil's Inequality Coefficient
%Data are observed and predicted values of deer kill;
obs = [218 206 139 127 113 76 56 58 47 33]';
pred = [241 196 134 137 74 78 59 55 59 51]';
%plot(obs,pred,'ro')
%axis([0 250 0 250])
%xlabel ('Observed Deer Kill')
%ylabel ('Predicted Deer Kill')
e = obs-pred;
esq = e.^2;
ebar = mean(esq);
eroot = sqrt(ebar);
obssq = obs.^2;
obsbar = mean(obssq);
obsroot = sqrt(obsbar);
predsq = pred.^2;
predbar = mean(predsq);
predroot = sqrt(predbar);
U = eroot./(obsroot + predroot)
```

The Theil m-file returns the value of U:

```
U =
0.0667
```

This value is close to zero, which indicates close to zero differences between predicted and observes values, or nearly perfect predictions. This result confirms the results of the statistical tests in the previous examples.

10.4.5 ANALYSIS OF VARIANCE

Analysis of variance (ANOVA) is used to test for differences between predicted and observed values when they can be grouped by factors. The MATLAB function **anova1** performs a one-way ANOVA for comparing the means of two or more groups of data. The MATLAB syntax is:

```
[p,anovatab,stats] = anova1(x,group,displayopt)
```

If **x** is a matrix, **anova1** treats each column as a separate group, and determines whether the population means of the columns are equal. This form of **anova1** is appropriate when each group has the same number of elements (balanced ANOVA). **group** can be a character array or a cell array of strings, with one row per column of **x**, containing the group names. Enter an empty array ([ ]) or omit this argument if you do not want to specify group names. If **x** is a vector, **group** must be a vector of the same length, or a string array or cell array of strings with one row for each element of **x**. **x** values corresponding to the same value of **group** are placed in the same group. **displayopt** can be "on" (the default) to display figures containing a standard one-way ANOVA table and a box plot, or "off" to omit these displays. The box plot option produces a box plot of the data in **x**. If **x** is a matrix, there is one box per column, and if **x** is a vector, there is just one box. On each box, the central mark is the median, the edges of the box are the 25th and 75th percentiles, the whiskers extend to the most extreme data points the algorithm considers not to be outliers, and the outliers are plotted individually. **anova1** uses the values in **group** as labels for the box plot of the samples in **x**, when **x** is a matrix.

The standard ANOVA table has columns for the sums of squares, degrees of freedom, mean squares (SS/df), *F* statistic, and *p* value.

The output from **[p,anovatab,stats]** = **anova1(...)** includes the *p* value for the null hypothesis that the means of the groups are equal; the ANOVA table values as the cell array **anovatab**; and an additional structure of statistics, **stats**, useful for performing a multiple comparison of means with the **multcompare** function.

If the experimental design has a second factor in addition to the observed and predicted grouping, a two-way ANOVA will test for the significance of the second factor using the MATLAB function **anova2**. The syntax is:

```
[p,table,stats] = anova2(x,reps,displayopt)
```

where the output is the same as for **anova1**. **anova2** compares the means of two or more columns and two or more rows of the sample in matrix **x**. The data in different columns represent changes in one factor; the data in different rows represent changes in the other factor. If there is more than one observation per row-column pair, the argument **reps** indicates the number of observations per "cell." A cell contains **reps** number of rows. As with **anova1**, **displayopt** can be "on" (the default) to display the ANOVA table, or "off" to skip the display. You can copy a text version of the ANOVA table to the clipboard by using the **Copy Text** item on the **Edit** menu in the Figure Window.)

10.4.6 KRUSKAL-WALLIS NONPARAMETRIC ANOVA

The nonparametric equivalent of the one-way ANOVA is the Kruskal-Wallis nonparametric ANOVA. Assumptions of the Kruskal-Wallis test are that all sample populations have the same continuous distribution, apart from a possibly different location, and all observations are mutually independent. The classical one-way ANOVA test replaces the first assumption with the stronger assumption that the populations have normal distributions.

The MATLAB function **kruskalwallis** performs a nonparametric one-way ANOVA for comparing the medians or means of two or more groups of data. The MATLAB syntax is:

```
[p,anovatab,stats] = kruskalwallis(x,group,'displayopt')
```

If **x** is a matrix, **kruskalwallis** treats each column as a separate group and determines whether the population medians of the columns are equal. This form of input is appropriate when each group has the same number of elements (balanced). **group** can be a character array or a cell array of strings, with one row per column of **x**, containing the group names. Enter an empty array ([ ]) or omit this argument if you do not want to specify group names. If **x** is a vector, **group** must be a vector of the same length, or a string array or cell array of strings with one row for each element of **x**. **x** values corresponding to the same value of **group** are placed in the same group. **displayopt** can be "on" (the default) to display figures containing a boxplot and the Kruskal-Wallis version of a one-way ANOVA table, or "off" to omit these displays.

The output from **[p,anovatab,stats] = kruskalwallis(...)** includes the p value for the null hypothesis that the medians of the groups are equal, the ANOVA table values as the cell array **anovatab**, and an additional structure of statistics, **stats**, useful for performing a multiple comparison of means with the **multcompare** function.

Example 10.4

In this example, we compare predicted with observed litter biomass values (Table 10.4) from a model by Dixon et al. (1978a, b).

The observed and predicted values are compared using **anova2** in the m-file, **anova**. The first noncomment line defines the seasons numerically in the order of Table 10.4. The next three commands create a matrix **x** and the columns representing observed and predicted litter values. We next plot the data (Figure 10.3) and set the x axis ticks and tick labels before running the analysis.

```
%m-file anova
%Analysis of Variance of difference between predicted
%and observe litter values
%column 1 is observed, column 2 is predicted,
%rows are seasons: Spring Summer Fall Winter
Season = [1; 2; 3; 4]
x = [1330.8 1330.8
     1162.0 1217.8
     1481.0 1432.8
     1605.5 1567.8]

observed = [x(:,1)]
```

TABLE 10.4
Observed and Predicted Litter Values for Four Seasons of the Year

Season	Observed	Predicted
Spring	1330.8	1330.0
Summer	1162.0	1217.8
Fall	1481.0	1432.8
Winter	1605.5	1567.8

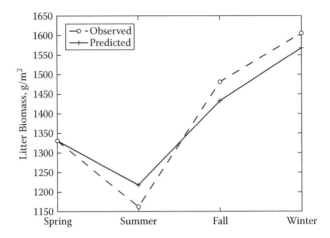

FIGURE 10.3   Predicted and observed litter biomass in spring, summer, fall, and winter.

ANOVA Table

Source	SS	df	MS	F	Prob>F
Columns	113.3	1	113.3	0.1	0.7699
Rows	173373.9	3	57791.3	52.29	0.0043
Error	3315.8	3	1105.3		
Total	176803	7			

FIGURE 10.4   Analysis of variance table for the test of differences in litter biomass among seasons.

```
predicted = [x(:,2)]
group = {'OBSERVED', 'PREDICTED'};
plot(Season,observed,'r--o')
hold on
plot(Season,predicted,'b-+')
hold off
set(gca,'XTick',1:1:4);
set(gca,'XTickLabel',{'Spring','Summer','Fall','Winter'})
ylabel ('Litter Biomass, g/m^2')
legend('observed','predicted',0)
[p,tbl,stats] = anova2(x,1)
```

The output of the **anova** m-file includes two figures: the plot of the observed and predicted values (Figure 10.3) and the ANOVA table (Figure 10.4).

The results of the **anova2** function shown in the output table indicate that we were not able to show a significant difference between observed and predicted litter values ($F = 0.0125$, $P > F = 0.7699$). There was a significant difference in litter biomass among seasons ($F = 52.3$, $P > F = 0.0043$). The results in the Command Window include the groups, data, statistics, and ANOVA table.

```
Season =
     1
     2
     3
     4
```

```
x =
   1.0e+003 *
      1.3308      1.3308
      1.1620      1.2178
      1.4810      1.4328
      1.6055      1.5678

observed =
   1.0e+003 *
      1.3308
      1.1620
      1.4810
      1.6055

predicted =
   1.0e+003 *
      1.3308
      1.2178
      1.4328
      1.5678

p =
      0.7699      0.0043

tbl =
   Columns 1 through 5
      'Source'       'SS'                  'df'      'MS'                  'F'
      'Columns'      [    113.2513]        [ 1]      [    113.2513]        [ 0.1025]
      'Rows'         [1.7337e+005]         [ 3]      [5.7791e+004]         [52.2867]
      'Error'        [3.3158e+003]         [ 3]      [1.1053e+003]         []
      'Total'        [1.7680e+005]         [ 7]      []                    []
   Column 6
      'Prob>F'
      [0.7699]
      [0.0043]
             []
             []

stats =
         source: 'anova2'
        sigmasq: 1.1053e+003
       colmeans: [1.3948e+003 1.3873e+003]
           coln: 4
       rowmeans: [1.3308e+003 1.1899e+003 1.4569e+003 1.5867e+003]
           rown: 2
          inter: 0
           pval: NaN
             df: 3
```

EXERCISES

1. Use linear regression to test the predicted litter biomass against the observed data in Table 10.4 using the MATLAB function **regress**. Plot the predicted over the observed data. What is the regression coefficient? Is it significantly different from 1.0? What is the r-square statistic to four significant digits?

2. Use analysis of variance to test for significant differences between predicted and observed deer kill data in Table 10.3. Plot predicted and observed deer kill over days. What is the F value and the probability of $F > 0.05$ for differences among days? Between predicted and observed?

REFERENCES

Burns, L. A. 2001. *Probabilistic Aquatic Exposure Assessment for Pesticides, I: Foundations.* EPA/600/R-01/071. Research Triangle Park, NC: U.S. Environmental Protection Agency.

Dixon, K. R., R. J. Luxmoore, and C. L. Begovich. 1978a. "CERES—A Model of Forest Stand Biomass Dynamics for Predicting Trace Contaminant, Nutrient, and Water Effects. I. Model Description." *Ecology Modelling* 5:17–38.

Dixon, K. R., R. J. Luxmoore, and C. L. Begovich. 1978b. "CERES—A Model of Forest Stand Biomass Dynamics for Predicting Trace Contaminant, Nutrient, and Water Effects. II. Model Application." *Ecology Modelling* 5:93–114.

Hermann, C. 1967. "Validation Problems in Games and Simulation with Special Reference to Models of International Politics." *Behavioral Science* 12:216–230.

Jacobs, D., and K. R. Dixon. 1982. "A Queuing Model of White-Tailed Deer Harvest." *Journal of Wildlife Management* 46:325–352.

Mankin, J. B., R. V. O'Neill, H. H. Shugart, et al. 1975. "The Importance of Validation in Ecosystem Analysis." In *New Directions in the Analysis of Ecological Systems, Simulation Councils Proceedings, Series 1(1),* ed. G. S. Innis, 63–72. La Jolla, CA: Simulation Councils.

Mihram, G. A. 1971. "Some Practical Aspects of the Verification and Validation of Simulation Models." *Operational Research Quarterly* 23:17–29.

Naylor, T. H., and J. M. Finger. 1967. "Verification of Computer Simulation Models." *Management Science* 14(2): B-92–B-101.

Petroski, H. 1992. *To Engineer Is Human: The Role of Failure in Successful Design.* New York: Vintage Books.

Platt, J. R. 1964. "Strong Inference." *Science* 146:347–353.

Theil, H. 1970. *Economic Forecasts and Policy.* Amsterdam: North-Holland Publishing Co.

Van Horn, R. L. 1971. "Validation of Simulation Results." *Management Science* 17:247–258.

11 A Model to Predict the Effects of Insecticides on Avian Populations

11.1 PROBLEM DEFINITION

Insecticides are used to protect crops from damage from insects either by killing the insects or preventing their feeding on the crops. Birds feeding on insects in treated fields can be exposed to lethal doses of the insecticide. The objective of this modeling effort is to predict the amount of the organophosphate insecticide, chlorpyrifos, consumed by avian species in cornfields in the U.S. Midwest, and the level of mortality resulting from this exposure pathway. The system includes concentrations of chlorpyrifos in diet components of four avian species with different feeding behaviors. The four species are ring-necked pheasant (*Phasianus colchicus*), northern bobwhite (*Colianus virginianus*), red-winged blackbird (*Agelaius phoeniceus*), and house sparrow *(Passer domesticus)*. A similar model was used in the risk assessment conducted by Solomon et al. (2001).

11.2 MODEL DEVELOPMENT

Because individual birds are exposed to chlorpyrifos and may die as a result, we determined that the best modeling approach was an individual-based model. The model is stochastic in that feeding rates and mortality rates are expressed as random variables. The model, in conceptual terms, is simple since a single bird of a given species is simulated at a time. For each individual, the state variables are *body burden* and *mortality*. Body burden is determined by the amount of chlorpyrifos ingested, either as residues on food items or as chlorpyrifos granules, and the rate of excretion. Mortality is a function of body burden. The controlled input variables are the chlorpyrifos concentrations on food items (determined by spraying frequency and chlorpyrifos concentration of the spray) and the density of chlorpyrifos granules.

A simplified material flow diagram shows the arrangement of the subsystems and the flow of chlorpyrifos among them (Figure 11.1).

11.3 MODEL IMPLEMENTATION

In model implementation, we develop a quantitative expression for the system state variables, followed by quantitative descriptions of subsystem mechanisms such as ingestion and excretion, and the interrelationships among subsystem components. The subsystems then are combined into a simulation model of the whole system. The model is written using a difference equation with a time step of one hour as this represents a realistic time interval to mimic avian feeding behavior. Time also is tracked on a daily basis to account for the daily granule consumption. Variable and parameter units were checked for consistency as parameter values were obtained from different sources.

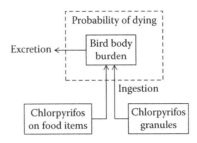

FIGURE 11.1   Material flow diagram for avian insecticide exposure model.

11.3.1   Model Description

The model is an individual-based mathematical model that consists of two parts: (1) an expression for the body concentration or dose for each individual in the population, and (2) an expression for the probability of mortality for the current dose. Each part of the model is stochastic to allow for Monte Carlo simulation. The change in body concentration of chlorpyrifos in an individual between time t and time $t + 1$ is the increase in dose from ingestion of chlorpyrifos granules or contaminated food minus the loss of chlorpyrifos from elimination, which can be described by the difference equation:

$$Q_{t+1} = Q_t + \sum_{i=1}^{n} I_{i,t}\alpha_{i,t} + G_t - \lambda Q_t \tag{11.1}$$

where
$\quad Q_t \quad$ = chlorpyrifos body burden at time t, mg·kg^{-1}
$\quad Q_{t+1}$ = chlorpyrifos body burden at time $t+1$, mg·kg^{-1}
$\quad I_{i,t} \quad$ = ingestion rate of chlorpyrifos in food item i at time t, mg·kg^{-1}·h^{-1}
$\quad \alpha_{i,t} \quad$ = proportion of total diet contributed by item i at time t
$\quad G_t \quad$ = consumption of chlorpyrifos granules at time t, mg·kg^{-1}·h^{-1}
$\quad \lambda \quad$ = elimination rate constant, h^{-1}

In the difference Equation (11.1), Q_{t+1} and Q_t, with units mg·kg^{-1} and a time step of 1 hour, together take the place of the derivative, dQ/dt, with units mg·kg^{-1}·h^{-1}. Therefore, the terms on the right side of Equation 11.1 (other than Q_t) also must have units mg·kg^{-1}·h^{-1}.

11.3.1.1   Ingestion in Food

The weight-specific mass ingestion rate of chlorpyrifos, $I_{i,t}$, (mg·kg^{-1}·h^{-1}) may be written as:

$$I_{i,t} = \frac{p_i C_i f_i v_i}{Wt} \tag{11.2}$$

where
$\quad p_i \quad$ = proportion of food item, i, consumed that is contaminated
$\quad C_i \quad$ = consumption rate of food item i, g·h^{-1}
$\quad f_i \quad$ = dry weight to wet weight conversion factor for food item i
$\quad v_i \quad$ = chlorpyrifos concentration in food item i, mg·kg^{-1}
$\quad Wt$ = consumer body weight, g

Food consumption rates. The amount of food consumed in grams per day (dry matter), C_i, was estimated using the power functions (Nagy 1987, USEPA 1993) that describe consumption as a function of body weight:

TABLE 11.1
Model Parameters

Parameter	Ring-Necked Pheasant	Northern Bobwhite	Red-Winged Blackbird	House Sparrow	Source
Proportion of plant food in diet, α_i	0.84	0.73	0.50	0.94	Martin et al. 1951
Body weight, Wt, in grams	1135	178	53	28	Dunning 1993
Proporton of time feeding in treated fields, p_f	0.15	0.01	0.16	0.24	Best et al. 1990; Frey et al. 1994
Proportion of time spent in a treated area, p_w					(See Section 11.4.1)
Spray	1.00	1.00	1.00	1.00	
Granule	0.19	0.19	0.19	0.19	
Mean number of granules consumed per day	20.88	6.99	2.63	22.00	(See Section 11.3.2)
Excretion rate constant, λ, d^{-1}	0.51	0.51	0.51	0.51	Bauriedel 1986;
LD$_{50}$, P_2, in mg/kg,	8.41	32.00	13.10	10.00	Hudson et al. 1984; Tucker and Crabtree 1970; Hill and Camardese 1984; Schafer 1972; Schafer et al. 1973
LD$_{50}$-LD$_{10}$, P_3, in mg/kg (percentage of P_2)	2.00 (23.8)	7.64 (23.9)	3.13 (23.9)	2.39 (23.9)	(See Section 11.4.3)

Source: Data from K. R. Solomon, J. P. Giesy, R. J. Kendall, L. B. Best, J. R. Coats, K. R. Dixon,, M. J. Hooper, E. E. Kenaga, and S. T. McMurry. 2001. "Chlorpyrifos: Ecotoxicological Risk Assessment for Birds and Mammals in Corn Agroecosystems." *Human Ecological Risk Assessment* 7:497–632. With permission from Taylor & Francis.

$$C_i = \begin{cases} 0.398\ Wt^{0.850} & \text{passerines} \\ 0.301\ Wt^{0.751} & \text{nonpasserines} \end{cases} \tag{11.3}$$

To obtain hourly consumption rates and convert the units to g·h^{-1}, C_i was divided by the number of hours per day spent feeding. The passerine function was used for all focus species except the ring-necked pheasant. Body weights were obtained from Dunning (1993) (Table 11.1).

Dry Weight to Wet Weight Conversion Factor. Because chlorpyrifos residues are based on mg of chlorpyrifos per kg (wet weight) of tissue and food consumption is based on dry weight, a factor to convert dry weight to wet weight is needed. The conversion factor, f_i, is a function of water content:

$$f_t = \frac{1}{1 - p_{H_2O}} \tag{11.4}$$

where p_{H_2O} is the proportion of water in the food item (unitless).

11.3.1.2 Consumption of Chlorpyrifos Granules

The equation for the contribution to body burden from ingestion of granules is similar to that for food ingestion (Equation [11.5]):

$$G_t = \frac{p \cdot Wg \cdot Vg \cdot Dg \cdot 1000 \cdot Ng_t}{Wt} \tag{11.5}$$

where

p = proportion of time spent in granule-treated areas
Wg = weight of granules, mg
Vg = granule insecticide concentration, mg·mg^{-1}
Wt = consumer body weight, g
Dg = dissipation rate of granular insecticide, h^{-1}
Ng_t = number of granules ingested

The 1000 in the numerator makes the units on the right side of the equation conform to those on the left side: mg·kg^{-1}·h^{-1}

Number of insecticide granules ingested. Consumption of insecticide granules, Ng_t, was assumed to be a Bernoulli random variable. The number of granules ingested was considered a Bernoulli trial in which a granule was ingested with probability p:

$$p(x) = \begin{cases} 1-p & \text{if } x = 0 \\ p & \text{if } x = 1 \\ 0 & \text{otherwise} \end{cases} \tag{11.6}$$

The assumption of a Bernoulli process applies at an hourly time step in the model. A daily summation of granules ingested using the Bernoulli distribution yields a binomial distribution.

Chlorpyrifos granule degradation. The integrated material balance equations for the degradation of chlorpyrifos from the granules, developed by Cryer and Laskowski (1994), were incorporated into the model. The amount of chlorpyrifos in the granule at time t, C_A, is dissipated by diffusion into the soil and volatilization into the atmosphere:

$$C_A = C_{A^o} e^{-(k_1+k_3)t} \tag{11.7}$$

where

k_1 = the rate constant for release of chlorpyrifos into the soil, and
k_3 = the rate constant for volatilization.

11.3.1.3 Avian Loss Rates

The primary mechanisms of chlorpyrifos removal from avian species are excretion and metabolism of absorbed chlorpyrifos and voiding of the chlorpyrifos granules. A single compartment elimination model was used to obtain elimination rate constants, λ, in Equation (11.1):

$$\lambda = \frac{-\log(a(1-p))}{t} \tag{11.8}$$

where a is the fraction of excreta that is chlorpyrifos (≤ 0.5), p is the fraction of dose excreted, and t is time between dosing and final sampling of excreta. The more conservative p value of 0.88 excretion fraction was used for an estimate of λ of 0.5116 day^{-1}.

Elimination of chlorpyrifos granules from gizzards (Fischer and Best 1995) also showed a negative exponential decrease to a plateau in both house sparrows and red-winged blackbirds.

11.3.1.4 Mortality

Mortality response function. The probability of mortality occurring in an individual is determined by the dose–response function in which mortality probability is a logistic function of dose or body concentration, Q (Section 5.1.4.3). The following form of the logistic function was used:

$$F(Q) = \frac{P_1}{1 + e^{(2.2/P_3)(P_2 - Q)}} \tag{11.9}$$

where

$F(Q)$ = probability of mortality at dose Q
P_1 = maximum probability of mortality
P_2 = LD_{50}
P_3 = difference between LD_{10} and LD_{50}
Q = dose or body concentration

Mortality probability function. To determine the quantal response (i.e., whether or not mortality occurs), a random number generator was used to obtain a sample from a uniform distribution, $U(0,1)$. If the value of the random variable is less than or equal to $F(Q)$, mortality is assigned to the individual. A population response is obtained by simulating many individuals.

11.3.2 MODEL STRUCTURE VALIDATION

At this step in the simulation process, a test of model validity involves examination of the conceptual model, the logical flowchart, and the model structure for logical errors or inconsistencies. Where comparable data exist, we can compare subsystem dynamics with these data. Consumption of chlorpyrifos granules was assumed to be a Bernoulli random variable. The number of granules actually consumed in a given day of the simulation was assumed to follow a binomial probability mass function, which is equivalent to a sum of Bernoulli trials. The assumption of a binomial distribution was based on frequency histograms of granule counts in gizzards using data from Fischer and Best (1995) and Gionfriddo and Best (1996) (Figure 11.2). We used the MATLAB function, **binopdf**, to generate the binomial distribution in making the comparison with the daily granule consumption data (Figure 11.2):

```
% m-file binomial
% program to compare binomial pdf with experimental data
x = [0 1 2 3 4 5 6];              % classes of number of granules
y1 = binopdf(x,25,.018);          % generate binomial pdf
y1 = y1';                         % transpose y1
y2 = [.60 .18 .10 .08 .03 .02 .01]; % data from Fisher and Best
y2 = y2';                         % transpose y2
y = [y1 y2];                      % form matrix of column vectors,
                                  % y1 and y2
figure
bar(x,y)                          % generate bar graph
xlabel('Number of Granules in Gizzard')
ylabel('Proportion of Birds')
legend1 = legend({'predicted','observed'},...
          'Position',[0.5901 0.5647 0.1607 0.1889]);
```

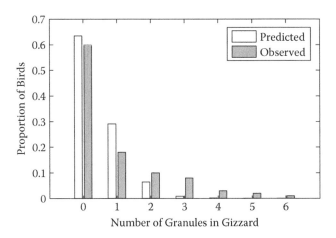

FIGURE 11.2   Observed and predicted binomial distribution of daily ingestion of insecticide granules. (Adapted from D. L. Fischer and L. B. Best. 1995. "Avian Consumption of Blank Pesticide Granules Applied at Planting to Iowa Cornfields." *Environmental Toxicology and Chemistry* 14, 9:1543–1549.)

11.3.3 PROGRAMMING THE COMPUTER CODE

Development of the computer code includes selection of a simulation language, construction and verification of a logical flowchart, and writing the program code. We used MATLAB as the simulation language. As there is a need for a daily loop and an hourly loop nested within the daily loop, to describe the model dynamics, we begin with a somewhat detailed programming flowchart (Figure 11.3).

11.4 DATA REQUIREMENTS

The parameter values used in the model were obtained from studies published in either the open literature or in technical reports provided by DowElanco. Initial values for all state variables of body burden and mortality are assumed to be zero. We assumed a starting population of 100 for each species. All parameter values were obtained from studies on the four focus species or similar species.

11.4.1 INGESTION

11.4.1.1 Proportion of Components in Diet

The four focus species show a varying diet, with the proportion of plant food, α_i, ranging between 50 and 94% (Table 11.1). The percentage in the summer diet was used because that is most representative of the exposure period from spray applications. The balance of the diet consists of animal components.

11.4.1.2 Granule Consumption Rate

Probabilities of granule consumption used in the Bernoulli distribution were estimated from data on grit consumption in Iowa cornfields (Fischer and Best 1995, Gionfriddo and Best 1996) and granule consumption by house sparrows (Best and Gionfriddo 1994). The first method used the mean number of grit particles in gizzards from Gionfriddo and Best (1996), multiplied this number by 4.2 to convert the estimate to granules consumed per day (Solomon et al. 2001), and then divided this number by 6 to adjust for the difference between ingestion of clay and silica granules. The second method used estimates of clay granule consumption in house sparrows from Best and Gionfriddo (1994). For this species, the estimated average daily consumption of clay granules was 21±21.9. For

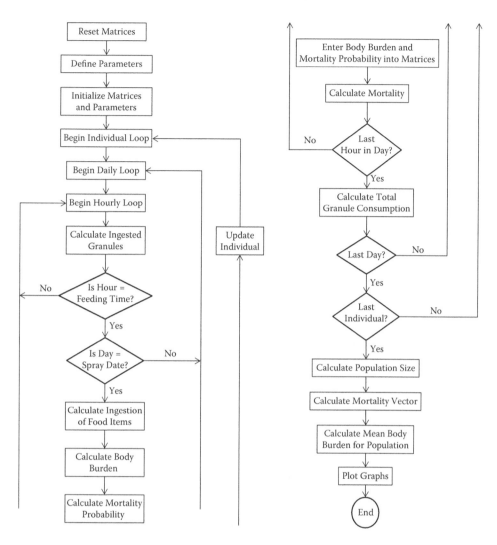

FIGURE 11.3 Programming flowchart for avian insecticide exposure model.

the other focus species, we multiplied this estimate by the ratio of the mean number of grit particles in the gizzards of the focus species to that of the house sparrow. The greater of the estimates from these two methods was used in the simulations. Probabilities were estimated as the number of granules consumed per hour.

11.4.1.3 Time Spent in Treated Areas

The proportion of consumed food items that are contaminated with chlorpyrifos, p_i, will depend upon the relative time spent in treated areas compared to untreated areas. The untreated areas include both untreated areas surrounding treated agricultural fields and, in the case of banded applications, the areas between bands. The proportion of food items contaminated, p_i, is the product of the proportion of time feeding in a treated field, p_f, and the proportion of time feeding in treated areas within the field given that the bird is feeding in the field, p_w.

Edge vs. field. The time spent feeding within a field and adjacent to the field can be estimated from observations of the number of birds feeding in each area. Assuming that the location of feeding over a period of time is a random process, the proportion of time spent feeding in a field will be equal to the proportion of the total number of birds observed in the field. Data from Iowa and Illinois cornfield studies (Best et al. 1990, Frey et al. 1994) (See Solomon, et al. 2001, Table 21) were used to obtain estimates of p_f. The highest reported field-use percentage was used.

Band vs. nonband. Chlorpyrifos granules are applied in bands approximately 18.0 cm wide and 76.2 cm apart. On an area basis, the proportion of time spent in a treated area is 0.19 (Fischer and Best 1995). Therefore, a value of p_w of 0.19 was used for granule consumption. Although spray treatments with 4E sometimes are applied in bands, a continuous coverage was assumed. Therefore, p_w was 1.0 for other food items.

11.4.1.4 Residues in Diet Components

Dietary components. Residue concentrations depend upon the application rate. The parameter values are based upon a maximum application rate of 1.7 kg ·ha⁻¹. The concentration of chlorpyrifos in the plant component (parameter, v_i) was taken from data on seed residues (Solomon et al. 2001, Table 11). The maximum value of 13.5 mg·kg⁻¹ was used in the model. The mean of three sampling locations was used in which samples were obtained immediately following application date when residues were at a maximum. For the insect component of the diet, residue data were obtained on invertebrates collected from corn fields treated with sprayable chlorpyrifos formulation (Frey et al. 1994). The maximum value of 7.7 mg·kg⁻¹ (Solomon et al. 2001, Table 17) was used in the model. Each residue value was treated as a normally distributed random variable. The random variates $N(0,1)$ were generated using the MATLAB function **randn**.

Weight of granules, Wg. Median chlorpyrifos granule weight is 0.064 mg (Hill and Camardese 1984, Table 1).

Chlorpyrifos concentration in granules, Vg. Chlorpyrifos concentration in granular formulation is 15%. Therefore, the value of Vg is 0.15.

Chlorpyrifos granule degradation. The rate constants for diffusion into the soil, k_1, and volatilization into the atmosphere, k_3, in Equation (11.7) have a combined value of 0.0141 (Cryer and Laskowski 1994).

Dry Weight to Wet Weight Conversion Factor. Because chlorpyrifos residues are based on mg of chlorpyrifos per kg (wet weight) of tissue and food consumption is based on dry weight, a factor to convert dry weight to wet weight is needed. The conversion factor, f_i, is a function of water content. The water content of the three food items used in the model were 0.10, and 0.50, for seeds, and insects, respectively.

Dissipation from diet components following application. Chlorpyrifos dissipates rapidly from plant surfaces following spraying. In a study of corn sprayed at a rate of 1.12 kg/ha, dissipation resulted primarily from volatilization (McCall et al. 1985). After two days, 79.3% of chlorpyrifos had volatilized. The mean half-life of dissipation from plants and seeds is 3.9 days (see Solomon et al. 2001, Table 9). The dissipation rate constant was estimated by assuming an exponential decay function:

$$C_t = C_0 e^{-k \cdot t}$$

$$\frac{C_t}{C_0} = e^{-k \cdot t}$$

$$0.5 = e^{-k \times 3.9} \tag{11.10}$$

$$\ln(0.5) = -k \times 3.9$$

$$-0.1777 = -k$$

The estimated rate constant is rounded to 0.18 day^{-1} or 0.0074 h^{-1}. Similar rates were reported in other studies summarized by Racke (1993, Table 11). The rate of elimination of chlorpyrifos from the animal components was estimated from chlorpyrifos data on leatherjackets (*Tipula* spp.) from a study by Clements and Bale (1988) in Great Britain. From a peak value 1.17 mg/kg, residues dropped to 0.64 mg/kg after four days. Assuming a negative exponential elimination function, the estimated rate constant is 0.15 day^{-1}:

$$C_t = C_0 e^{-k \cdot t}$$

$$0.64 = 1.17 e^{-k \cdot t}$$

$$\frac{0.64}{1.17} = e^{-k \times 4.0} \tag{11.11}$$

$$\ln(0.5470) = -k \times 4.0$$

$$-0.1508 = -k$$

This value was rounded to 0.15 d^{-1} or 0.0063 h^{-1}.

11.4.2 AVIAN LOSS RATES

The primary mechanisms of chlorpyrifos removal from avian species are excretion and metabolism of absorbed chlorpyrifos and voiding of the chlorpyrifos granules. The total loss rates from excretion and metabolism were estimated from a study in hens by Bauriedel (1986) in which laying chicken hens were exposed to 20 ppm dietary chlorpyrifos. During a 10-day exposure period, 88–95% of the dose was excreted via droppings. Less than 5% of the excreted dose was chlorpyrifos; the rest had been metabolized, primarily as 3,5,6-trichloro-2-pyridinol (TCP). The dose from TCP or other metabolites was not included in the model because TCP was found to be less toxic to birds than parent chlorpyrifos (Marshall and Roberts 1978). The value of the elimination rate constant, λ, was estimated as 0.5116 day^{-1} using Equation (11.8).

11.4.3 MORTALITY

Mortality response function. Data on LD$_{50}$ and slope (Solomon et al. 2001, Table 28) were used to estimate the parameters in the logistic function (Equation [11.9]). The parameter P_2 is exactly the LD$_{50}$ value. Where more than one value was reported, we used the lowest value to assure a conservative set of predictions. The slope is the parameter derived from probit analysis. It is important to have both LD$_{50}$ and slope values to define the shape of the dose–response curve. Only the house sparrow and the northern bobwhite, however, had both LD$_{50}$ and slope estimates. Initial estimates of P_3 were made for these two species using the LD$_{50}$ and slope values. There is a one-to-one relationship between the slope and P_3. As the slope increases, the range between LD$_{50}$ and LD$_{10}$, and therefore the value of P_3, decreases. To obtain P_3 from the slope, we use the relation (Hill and Camardese 1986, 145):

$$\log LD_k = \log LD50 + \frac{(\text{probit } k - \text{probit } 5)}{b} \tag{11.12}$$

where
   k = the kth proportional response, e.g., the 10th percentile or probit 1
   b = slope

For example, the LD_{50} and slope for northern bobwhite are 32.0 and 4.60, respectively. To obtain the parameter P_3, we calculate:

$$\log LD10 = \log LD50 + \frac{(\text{probit } 1 - \text{probit } 5)}{b}$$

$$= \log(32) + \frac{3.7184 - 5.000}{4.60}$$

$$= 1.2265$$

$$LD10 = 16.8438 \tag{11.13}$$

$$P_3 = LD50 - LD10$$

$$= 32.0000 - 16.8438$$

$$= 15.1542$$

There also is a relationship between the parameters P_2 and P_3 (Figure 5.5). The parameter P_3 has to be less than P_2 if we define the dose–response curve at the point (0,0). In other words, there has to be zero mortality probability at a zero dose. As P_2 decreases, the value of P_3 also decreases and then levels off at about 45% of the P_2 value. Plotting the resultant dose–response curve showed that P_3 had to be revised downward to obtain a (0,0) point. This procedure for estimating P_3 was repeated for the other species, beginning with an estimate of 45% of the P_2 value. This value was adjusted until the estimated mortality probability was ~0.0001 at zero dose. The estimated P_3 value as a percentage of P_2 values was nearly identical for all four species (Table 11.1).

Mortality probability function. To determine the quantal response (i.e., whether or not mortality occurs), the MATLAB function, **rand**, was used to obtain a sample from a uniform distribution, $U(0,1)$. If the value of the random variable was less than or equal to $F(Q)$, mortality was assigned to the individual. A population response was obtained by simulating many individuals.

11.5   MODEL VALIDATION

No mortality data were collected to test the model. Therefore, we used model validation tests for a "first-time" model (Section 9.2). One test, the variable-parameter validity test, was conducted on the ingestion rate of chlorpyrifos granules (Section 11.3.2). Model parameters were compared with their real-world counterparts obtained from laboratory experiments. Internal validity also was tested where stochastic variables were checked for low variance of outputs. None of the stochastic variables was found to obscure changes in output resulting from changes in controlled or environmental variables.

11.6   DESIGN SIMULATION EXPERIMENTS

Simulations were run for each of the four focus species in Table 11.1, with dose (body burden) and population size as state variables. In other words, both dose and survival were followed over time. Although the model captures the essence of what is known about avian feeding behavior in and around cornfields, several conservative assumptions were made that assured that these simulations produced estimates of population mortality that were not likely to be exceeded in field situations:

1. Where data were missing for a given species, a conservative estimate based on values for other avian species was used.
2. The maximum mean values of chlorpyrifos residues were used for the parameter v_i.

3. The process contributing to the greatest rate of chlorpyrifos degradation on the granule, advection during rainfall events, was not used in the model.
4. The elimination rate from birds was based on the lowest reported value of percentage reduction in body burden.
5. For those species for which no LD_{50} data were available, the lower 99% confidence limit (CL) of the distribution of LD_{50} values was used.
6. The possible behavior of avoiding chlorpyrifos granules was not included in the model.

The exposure scenario for all simulations was that the start time of the simulation was the application date of chlorpyrifos granules at plant, followed by a single spray application 60 days later. This is another conservative aspect of the model because only about 5% of the cornfields are sprayed in a given year. The simulations, then, represent the worst case of chlorpyrifos treatments in cornfields. Any predicted mortality would not be representative of the entire Midwestern corn agroecosystem.

11.7 ANALYZE RESULTS OF SIMULATION EXPERIMENTS

Simulations were run for the four focus species: ring-necked pheasant, northern bobwhite, red-winged blackbird, and house sparrow, with five replicates per species. Each run simulated a population of 100 individuals. We examined both differences in body burden and mortality among species.

11.7.1 PREDICTED DOSE

Because there are so many parameters in the model that affect body burden, or dose, it is difficult to predict a priori which species would have the highest concentrations of chorpyrifos. Diet, body weight, proportion of time exposed to the insecticide, and number of granules consumed, all impact the dose. Simulation can predict the dose, allowing for the set of parameter values for each species. For each focus species, Monte Carlo simulations were run to obtain a population mean and 95% confidence interval of dose (Figures 11.4 to 11.7). The dose contribution from granule ingestion was significantly less than that from ingestion of spray application residues. The body burden pattern was the same for all four species with an initial increase in concentration from granule consumption followed by a significant increase at day 60 (hour 1440) from the spray application. All body burdens approached zero after the 120-day exposure scenario.

11.7.1.1 Ring-Necked Pheasant

The predicted dose was relatively low, with a peak of about 0.07 mg/kg, primarily a result of the high body weight. The contribution to exposure from 4E spray application is significantly greater than that from 15G granule application, even though pheasants had the second highest granule consumption rate (Figure 11.4).

11.7.1.2 Northern Bobwhite

The predicted dose was the lowest of the four species and slightly lower than that of pheasants, primarily because of the least amount of time spent in fields ($p_f = 0.01$). The lower number of granules consumed per day also contributed to a lower dose. The higher body weight led to a lower body burden than either red-wing blackbirds or house sparrows (Figure 11.5).

11.7.1.3 Red-Winged Blackbird

The predicted dose in the red-winged blackbird was the second highest of the four species. At a peak of about 0.14 mg/kg, it was 10 times greater than in the northern bobwhite. This resulted from a lower body weight and fewer numbers of granules consumed. The pattern was similar to that of the house sparrow, although the dose is lower as a result of the shorter amount of time spent in the field (16%) (Figure 11.6).

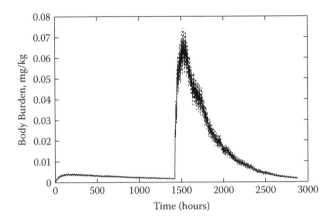

FIGURE 11.4   Mean (solid line) and 95% confidence intervals (dotted lines) of chlorpyrifos body burden in ring-necked pheasant.

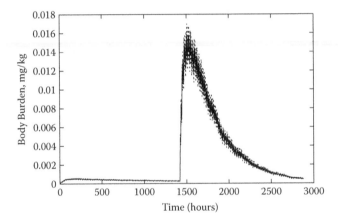

FIGURE 11.5   Mean (solid line) and 95% confidence intervals (dotted lines) of chlorpyrifos body burden in northern bobwhite.

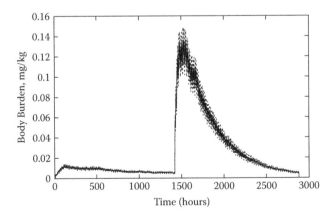

FIGURE 11.6   Mean (solid line) and 95% confidence intervals (dotted lines) of chlorpyrifos body burden in red-winged blackbird.

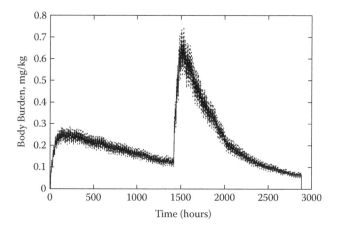

FIGURE 11.7   Mean (solid line) and 95% confidence intervals (dotted lines) of chlorpyrifos body burden in house sparrow.

11.7.1.4   House Sparrow

The house sparrow had the highest predicted dose of the four species. This result was caused by the highest rate of granule consumption, the lowest body weight, the highest proportion of time feeding in treated fields, and the highest proportion of plant food in the diet. There was a relatively higher contribution from granules compared to spray application because of the high rate of granule consumption (Figure 11.7).

11.7.2   Predicted Mortality

Each simulation also predicted mortality for each dose trace. Mortality was subtracted from the population of 100 individuals to estimate probability of survival. The model predicted mortality for all species, given the conservative assumptions included in the model. Once the body burden, or dose, is predicted, the level of mortality is still difficult to predict without simulating the exposure scenario because mortality will be affected by the parameters P_2 and P_3. For example, the house sparrow and the pheasant have the lowest LD_{50} values, but pheasants were predicted to have the lowest mortality whereas the house sparrow predicted mortality was the highest (Table 11.2).

TABLE 11.2
Predicted Mortality in Five Simulation
Experiments for Four Avian Species Exposed
to Chlorpyrifos Applications

Replicate	Pheasant	Northern Bobwhite	Red-Winged Blackbird	House Sparrow
1	17	23	26	27
2	24	29	32	31
3	22	25	21	33
4	23	31	23	39
5	25	22	37	33
Means	22.20	26.00	27.80	32.60

To compare the mean mortality among species, we ran the MATLAB **anova1** function in the m-file **mortalityanova**. We also ran individual mean comparisons using the **c = multcompare(stats)** command.

```
%m-file mortalityanova
%Analysis of Variance of difference in mortality among bird species
%column 1 is pheasant, column 2 bobwhite, column 3 is red-wing,
%column 4 is sparrow
%rows are replicates
Species = [1; 2; 3; 4];
X = [17 23 26 27
      24 29 32 31
      22 25 21 33
      23 31 23 39
      25 22 37 33];

pheasant = X(:,1);
bobwhite = X(:,2);
redwing = X(:,3);
sparrow = X(:,4);
GROUP = {'Pheasant','Bobwhite','Red-wing','Sparrow'};

[p,tbl,stats] = anova1(X,GROUP)
ylabel ('Mortality Percentage')
c = multcompare(stats)
```

The output from the m-file is displayed in the Command Window:

```
p =
      0.0213

tbl =
      'Source'      'SS'          'df'      'MS'          'F'         'Prob>F'
      'Columns'     [279.7500]    [ 3]      [93.2500]     [4.2775]    [0.0213]
      'Error'       [348.8000]    [16]      [21.8000]     []          []
      'Total'       [628.5500]    [19]      []            []          []

stats =
      gnames: {4x1 cell}
           n: [5 5 5 5]
      source: 'anova1'
       means: [22.2000 26 27.8000 32.6000]
          df: 16
           s: 4.6690

c =
      1.0000    2.0000    -12.2485    -3.8000     4.6485
      1.0000    3.0000    -14.0485    -5.6000     2.8485
      1.0000    4.0000    -18.8485   -10.4000    -1.9515
      2.0000    3.0000    -10.2485    -1.8000     6.6485
      2.0000    4.0000    -15.0485    -6.6000     1.8485
      3.0000    4.0000    -13.2485    -4.8000     3.6485
```

FIGURE 11.8   Analysis of variance comparing mean mortality in four avian species exposed to chlorpyrifos applications.

The **p** value of 0.0213 is the probability that the test statistic **F** would occur by chance under the null hypothesis that all species samples are drawn from populations with the same mean. We can assume that there is at least one significant difference between two means at the 0.05 level (0.0213 < 0.05). The **stats** output includes a vector **n** with the number of runs for each species, a vector **means** with the species mean mortalities, and the number of degrees of freedom **df**. The analysis of variance (ANOVA) table **tbl** is generated in the Command Window output as well as a figure (Figure 11.8).

The matrix **c** displayed in the Command Window shows the results of the multiple comparison tests. The first two columns of **c** are the individual mean comparisons. The fourth column is the difference between the two means, the third column is the lower 95% confidence limit of the difference, and the fifth column is the upper 95% confidence limit. For example, the first row compares pheasants (species 1) to bobwhites (species 2). The difference between these two species was found to be not significant since the confidence interval contains 0.0. These results are shown also in the figure generated from the **multcompare** procedure, which uses the **stats** data from the **anova1** procedure (Figure 11.9).

The multiple comparison of species mortality means in Figure 11.9 is an interactive plot. By clicking on a mean symbol for a species, the resulting comparisons are shown and any significant comparison is indicated by nonoverlapping confidence intervals. In the comparison shown in the figure, the mean mortality in pheasants is shown to be significantly different from the mean mortality in house sparrows, but not significantly different from the northern bobwhite or red-winged blackbird means.

The analysis of simulation experiments can tell us a lot about the system being simulated. The analysis of the relative effect that the dose response has on mortality showed that given the same dose and the same LD_{50}, the species with the lower value of the parameter P_3 (steepest slope) will have the lower mortality. This may seem counterintuitive; however, at the lower end of the dose–response function there is a higher probability of mortality with a larger P_3 value. The reverse is true of the probabilities at doses greater than the LD_{50}.

There is much uncertainty associated with the model parameters. An analysis of model sensitivity showed that parameter P_3 (the slope-related parameter) had the greatest effect on the model predictions (Figure 11.10).

Body weight has an initially high sensitivity but levels off between 2 and 5 after about 1000 hours. After initial transient behavior, the parameter P_2 increases and then levels off between −3 and −5. Granule consumption probability gradually decreases and then fluctuates about 0. The fact that the model is sensitive to parameters P_2 and P_3, suggests that both of these parameters, which define the dose response function, need to be estimated with great precision and accuracy. With the exception of the red-winged blackbird and the ring-necked pheasant, these parameters were estimated from the relationship between P_2 and P_3 because no other data were available. It is necessary to have both LD_{50} and slope values on the same species. This lends support for the

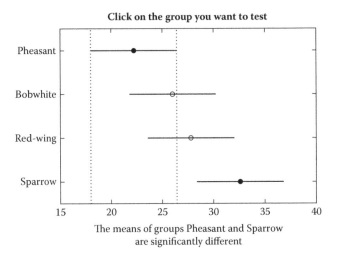

FIGURE 11.9 Multiple comparison of species mortality means.

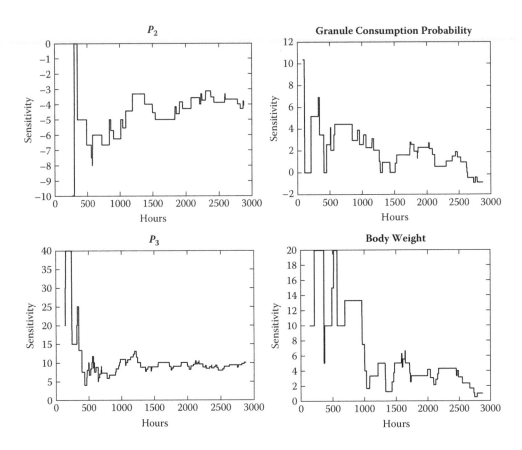

FIGURE 11.10 Sensitivity to selected parameters in the avian insecticide model.

reporting of both values in the publication of research. The model simulations and postsimulation analysis showed mortality in all species exposed to granular and spray applications of chlorpyrifos. These results are based on several conservative assumptions that tend to maximize both dose and response.

At this stage in the modeling process, additional data should be collected to better estimate parameters. One type of experiment would greatly improve the dose–response function. Now, toxicity tests are defined as either acute or chronic. Neither of these experiments provides the data to estimate the type of exposure experienced by birds in the real-world system. An acute test overestimates the dose, as the concentration of the toxicant exceeds what a bird would consume given a normal diet; however, mortality can be related directly to body burden. Chronic tests tend to underestimate the dose because the concentrations usually are not high enough to cause significant mortality. The added problem with chronic tests is that the response is related to toxicant concentration in food rather than body burden. From a modeling perspective, what is needed is an experiment that has a range of toxicant concentrations in food that will cause mortality and in which body burden can be measured. Given sufficient data, the model can provide improved estimates of dose and survival in avian species exposed to applications of insecticides in agricultural crops.

REFERENCES

Bauriedel, W. R. 1986. "Fate of ^{14}C - CPF Administered to Laying Hens." Unpublished report. Indianapolis, IN: DowElanco.

Best, L. B. and J. P. Gionfriddo. 1994. "House Sparrow Preferential Consumption of Carriers Used for Pesticide Granules." *Environmental Toxicology and Chemistry* 13:919–925.

Best, L. B., R. C. Whitmore, and G. M. Booth. 1990. "Use of Cornfields by Birds during the Breeding Season: The Importance of Edge Habitat." *American Midland Naturalist* 123:84–99.

Clements, R. O., and J. S. Bale. 1988. "The Short-Term Effects on Birds and Mammals of the Use of Chlorpyrifos to Control Leatherjackets in Grassland." *Annals of Applied Biology* 112:41–47.

Cryer, S. A., and D. A. Laskowski. 1994. "Chlorpyrifos Rate of Release from Lorsban®15G: Development of Algorithms for Use in Computer-Based Risk Assessment." Unpublished report. DowElanco, Indianapolis, IN: DowElanco.

Dunning, J. B. 1993. *CRC Handbook of Avian Body Masses.* Boca Raton, FL: CRC Press.

Fischer, D. L., and L. B. Best. 1995. "Avian Consumption of Blank Pesticide Granules Applied at Planting to Iowa Cornfields." *Environmental Toxicology and Chemistry* 14:1543–1549.

Frey L. T., D. A. Palmer, and H. O. Kruger. 1994. *LORSBAN Insecticide: An Evaluation of its Effects upon Avian and Mammalian Species on and around Corn Fields in Iowa.* Easton, MD and New Ulm, MN: MVTL Laboratories Inc. and Wildlife International, Ltd.

Gionfriddo, J. P., and L. B. Best. 1996. "Grit-Use Patterns in North American Birds: The Influence of Diet, Body Size, and Gender." *Wilson Bulletin* 108:685–696.

Hill, E. F., and M. B. Camardese. 1984. "Toxicity of Anticholinesterase Insecticides to Birds: Technical Grade versus Granular Formulations." *Ecotoxicological and Environmental Safety* 8:551–563.

Hill, E. F., and M. B. Camardese. 1986. *Lethal Dietary Toxicities of Environmental Contaminants and Pesticides to Controls.* Fish and Wildlife Technical Rept. 2. Washington, DC: U.S. Fish and Wildlife Service.

Hudson, R. H., R. K. Tucker, and M. A. Haegele. 1984. *Handbook of Toxicity of Pesticides to Wildlife,* 2nd ed. Resource Publication 153. Washington, DC: U.S. Fish and Wildlife Service.

Marshall, W. K., and J. R. Roberts. 1978. *Ecotoxicology of Chlorpyrifos.* Publ. No. NRCC 16079. Ottowa, Ontario, Canada: National Research Council of Canada.

Martin, A. C., H. S. Zim, and A. L. Nelson. 1951. *American Wildlife and Plants: A Guide to Wildlife Food Habits.* New York: Dover.

McCall, P. J., R. L. Swann, and W. R. Bauriedel. 1985. "Volatility Characteristics of Chlorpyrifos from Soil and Corn." Unpublished report. Indianapolis, IN: DowElanco.

Nagy, K. A. 1987. "Field Metabolic Rate and Food Requirement Scaling in Mammals and Birds." *Ecological Monographs* 57:111–128.

Schafer, E. W. 1972. "The Acute Oral Toxicity of 369 Pesticidal, Pharmaceutical, and Other Chemicals to Wild Birds." *Toxicology and Applied Pharmacology* 21:315–330.

Schafer, E. W, R. B. Brunton, N. F. Lockyer, and J. W. DeGrazio. 1973. "Comparative Toxicity of Seventeen Pesticides to the Quelea, House Sparrow, and Red-Winged Blackbird." *Toxicology and Applied Pharmacology* 26:154–157.

Solomon, K. R., J. P. Giesy, R. J. Kendall, L. B. Best, J. R. Coats, K. R. Dixon,, M. J. Hooper, E. E. Kenaga, and S. T. McMurry. 2001. "Chlorpyrifos: Ecotoxicological Risk Assessment for Birds and Mammals in Corn Agroecosystems." *Human Ecological Risk Assessment* 7:497–632.

Tucker, R.K., and D. G. Crabtree. 1970. *Handbook of Toxicity of Pesticides to Wildlife*. Resource Publication 84. Washington, DC: U.S. Fish and Wildlife Service.

U.S. Environmental Protection Agency (USEPA). 1993. *Wildlife Exposure Factors Handbook, Vol 1*. EPA report no. EPA/600/R-93/187a. Washington, DC: Office of Research and Development.

12 Case Study

Predicting Health Risk to Bottlenose Dolphins from Exposure to Oil Spill Toxicants

12.1 PROBLEM DEFINITION

The purpose of this case study is to simulate exposure of bottlenose dolphins (*Tursiops truncatus*) to chemicals in oil released into the marine environment from oil spills. The Deepwater Horizon oil spill is a reminder of the risks to wildlife, particularly marine mammals, of offshore oil drilling and releases from oil tankers. Three of the largest spills have been from leaking oil platforms. In addition to the Deepwater Horizon spill in summer 2010, with an estimated 4.3 million barrels of oil leaked into the Gulf of Mexico, there was the Ixtoc I spill off the Mexican gulf coast in 1979–80, with 3.5 million barrels, and the Nowruz Field Platform spill in the Persian Gulf in 1983 with 1.9 million barrels leaked. Since 1967 there have been at least 35 major spills from tanker ships (Table 12.1).

The constituents in crude oil vary considerably, depending upon the source. Most crude oil, however, is predominantly organic hydrocarbons, including polycyclic aromatic hydrocarbons, (PAHs) such as alkylated naphthalenes, phenanthrenes, and benzo(a)pyrene. These compounds may be taken up by marine mammals by ingestion of contaminated food, absorption by the skin, and inhalation of volatilized compounds. Little is known about the potential effects of exposure to PAHs by bottlenose dolphins, but animal studies have shown that certain PAHs can affect the hematopoietic and immune systems and can produce reproductive, neurologic, and developmental effects. When marine mammals (and other wildlife and human beings) are exposed to the mixture of PAHs found in crude oil, it is difficult to identify the effects of a single compound. It also is difficult to model the uptake and distribution of mixtures of PAHs by all exposure pathways at once. Experimental data are needed on the effects of both individual constituents and mixtures. In this case study, we limit simulations to the uptake and distribution of naphthalene from inhalation, although the model includes terms for skin absorption. In particular, we are interested in the effects of the different ventilation and cardiac output rates associated with dolphins diving below the surface.

The model is an example of a "first time" physiologically based toxicokinetic (PBTK) model designed to identify significant variables and parameters, and to develop hypotheses concerning the role that different types of swimming behavior have on the amounts of naphthalene in dolphin tissues.

12.2 MODEL DEVELOPMENT

We developed a PBTK model to predict naphthalene uptake and distribution in bottlenose dolphins. The model was based on a PBTK model for naphthalene inhalation in mice and rats (U.S. Department of Health and Human Services 2000). We also incorporated parts of a PBTK model of

TABLE 12.1
Major Oil Spills from Tanker Ships since 1967

Rank	Ship Name	Year	Location	Spill Size (tons)
1	*Atlantic Empress*	1979	Off Tobago, West Indies	287,000
2	*ABT Summer*	1991	700 nautical miles off Angola	260,000
3	*Castillo de Bellver*	1983	Off Saldanha Bay, South Africa	252,000
4	*Amoco Cadiz*	1978	Off Brittany, France	223,000
5	*Haven*	1991	Genoa, Italy	144,000
6	*Odyssey*	1988	700 nautical miles off Nova Scotia, Canada	132,000
7	*Torrey Canyon*	1967	Scilly Isles, UK	119,000
8	*Sea Star*	1972	Gulf of Oman	115,000
9	*Irenes Serenade*	1980	Navarino Bay, Greece	100,000
10	*Urquiola*	1976	La Coruna, Spain	100,000
11	*Hawaiian Patriot*	1977	300 nautical miles off Honolulu	95,000
12	*Independenta*	1979	Bosphorus, Turkey	95,000
13	*Jakob Maersk*	1975	Oporto, Portugal	88,000
14	*Braer*	1993	Shetland Islands, UK	85,000
15	*Khark 5*	1989	120 nautical miles off Atlantic coast of Morocco	80,000
16	*Aegean Sea*	1992	La Coruna, Spain	74,000
17	*Sea Empress*	1996	Milford Haven, UK	72,000
18	*Nova*	1985	Off Kharg Island, Gulf of Iran	70,000
19	*Katina P*	1992	Off Maputo, Mozambique	66,700
20	*Prestige*	2002	Off Galicia, Spain	63,000
35	*Exxon Valdez*	1989	Prince William Sound, Alaska, USA	37,000

Source: ITOPF (The International Tanker Owners Pollution Federation Limited). 2010. *Oil Tanker Spill Statistics: 2009.* ITOPF, London. http://www.itopf.com/information-services/data-and-statistics/statistics/documents/Statspack2009-FINAL.pdf (accessed 5/22/11).

nonane by Robinson (2000), which was based on a PBTK model of inhalation of styrene (Ramsey and Andersen 1984). A third model that we also examined was another PBTK model of naphthalene inhalation in mice and rats (Sweeney et al. 1996, Quick and Shuler 1999). All of these models were designed for inhalation exposure only (Figure 12.1).

The model consists of a calculation of the naphthalene concentration in each compartment for each individual in the population. The model, which is diffusion limited, contains compartments for arterial and venous blood, lung, liver, kidney, fat, skin, and "other" organs and tissues (Figure 12.1). The "other" organ compartment represents both slowly and rapidly perfused tissue (e.g., muscle, bone, heart, brain). Inhalation of naphthalene from ambient air concentrations takes place through a dolphin's "blowhole" to the alveolar space and then into the lung. Modeled uptake is dependent upon the ventilation rate, permeability of the tissue, and blood flow through the lung. Metabolism of naphthalene was assumed to take place primarily in the liver, but also in the lungs and skin. One metabolic pathway was used in both the lungs and skin, whereas in the liver, two pathways were used; one represented by Michaelis-Menten kinetics and the other by Hill kinetics. Dermal absorption takes place through naphthalene contact with the skin. Population responses were estimated by determining the response of many individuals in a population. The model is stochastic in that it contains random variables for naphthalene ambient air concentration. These random variables provide the capability to conduct stochastic simulations.

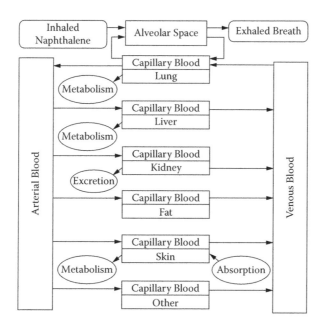

FIGURE 12.1 Flow diagram of a PBTK model for naphthalene inhalation and skin absorption. Adapted from U.S. Department of Health and Human Services. *NTP Technical Report on the Toxicology and Carcinogenesis Studies of Naphthalene (CAS No. 91-20-3) in F344/N Rats (Inhalation Studies).* National Toxicology Program, NTP TR 500, NIH Publication No. 01-4434 (Research Triangle Park, NC: National Toxicology Program, 2000).

12.3 MODEL IMPLEMENTATION

The model consists of a set of ordinary differential equations that were solved using fourth-order and fifth-order Runge-Kutta methods with default tolerances. The equations represent the dynamics of naphthalene as shown in Figure 12.1.

Naphthalene is inhaled from ambient air via the alveolar space (Equation [12.1]) into the lung capillary blood (Equation [12.2]). From the lung capillary blood, it goes either to arterial blood (Equation [12.4]) or to the lung tissue (Equation [12.3]) where it is metabolized. From the arterial blood, it is distributed to the liver (Equations [12.6] and [12.7]) and skin (Equations [12.8] and [12.9]) where it may be metabolized, or to other tissues (Equations [12.10] and [12.11]). Except for the lung capillary space, the effluent from all of the tissue capillary spaces goes to the venous blood compartment (Equation [12.5]). Naphthalene is transported via the venous blood to the lung capillary space (Equation [12.2]).

Symbols used to describe model equations are defined in Table 12.2. Parameters used in the model simulations are listed in Table 12.3.

12.3.1 DIFFERENTIAL EQUATIONS

The differential equations used in the model to describe absorption between the alveoli, lung capillaries, lung tissue, and blood are:

Alveolar space:

$$\frac{dAMT_{alv}}{dt} = Dose \cdot Q_{vent} + \frac{AMT_{lungcap}}{V_{lungcap}} \cdot \frac{Q_{vent}}{P_{air}} \cdot Perm - $$
$$\frac{AMT_{alv}}{V_{alv}} \cdot Q_{vent} \cdot Perm - \frac{AMT_{alv}}{V_{alv}} \cdot Q_{vent}$$

(12.1)

TABLE 12.2

Abbreviations and Symbols Used in Describing a PBTK Model for Naphthalene

Volumes, mL:

V_{alv}	Volume of alveolar space
$V_{lungcap}$	Volume of lung capillaries
V_{ven}	Volume of venous blood
V_{lung}	Volume of lung tissue
V_{art}	Volume of arterial blood
V_{liver}	Volume of liver tissue
$V_{skincap}$	Volume of skin capillaries
V_{skin}	Volume of skin tissue
$V_{tissuecap}$	Volume of kidney, fat, and "other" capillaries
V_{tissue}	Volume of kidney, fat, and "other" tissues

Concentrations, mg:

AMT_{air}	Amount in inhaled air
AMT_{alv}	Amount in alveolar air
AMT_{art}	Amount in arterial blood
AMT_{ven}	Amount in venous blood
$AMT_{lungcap}$	Amount in lung capillaries
AMT_{lung}	Amount in lung tissues
$AMT_{livercap}$	Amount in liver capillaries
AMT_{liver}	Amount in liver tissues
$AMT_{skincap}$	Amount in skin capillaries
AMT_{skin}	Amount in skin tissues
$AMT_{tissuecap}$	Amount in kidney, fat, and other tissue capillaries
AMT_{tissue}	Amount in kidney, fat, and other tissues

Flows, L/min:

Q_{vent}	Alveolar ventilation rate
Q_{total}	Total blood flow
Q_{liver}	Blood flow to liver
Q_{skin}	Blood flow to skin
Q_{tissue}	Blood flow to the kidney, fat, and other tissues

Partition Coefficients and Permeability Constant:

$Perm$	Capillary permeability constant
P_{air}	Blood:air partition coefficient
P_{lung}	Lung:blood partition coefficient
P_{liver}	Liver:blood partition coefficient
P_{skin}	Skin:blood partition coefficient
P_{tissue}	Kidney, fat, and other tissue:blood partition coefficients

Metabolism Rates:

$V_{max\ lung}$	Maximum lung enzymatic reaction rate (mg/min)
$V_{max\ liver1}$	Maximum liver Michaelis-Menten enzymatic reaction rate (mg/min)
$V_{max\ liver2}$	Maximum liver Hill enzymatic reaction rate (mg/min)
$V_{max\ skin}$	Maximum skin enzymatic reaction rate (mg/min)
K_{mlung}	Michaelis constant for lung enzymatic reaction (mg/liter blood)
$K_{mliver1}$	Michaelis constant for liver enzymatic reaction (mg/liter blood)
$K_{mliver2}$	Michaelis constant for lung Hill enzymatic reaction (mg/liter blood)
K_{mskin}	Michaelis constant for skin enzymatic reaction (mg/liter blood)
n	Hill constant

Source: Adapted from U.S. Department of Health and Human Services. 2000. *NTP Technical Report on the Toxicology and Carcinogenesis Studies of Naphthalene (CAS No. 91-20-3) in F344/N Rats (Inhalation Studies).* National Toxicology Program, NTP TR 500, NIH Publication No. 01-4434 (Research Triangle Park, NC: National Toxicology Program, 2000).

TABLE 12.3
Parameters Used in Model Simulations

Parameter Symbol	Parameter Description	Parameter Value
Physiological Parameters		
BW	Body weight (kg)	200.0
HR	Heart rate (beats/min)	90.0
SV	Stroke volume (L/beat)	0.0559
Vt	Tidal volume (L/breath)	5.0
f	Breathing frequency (breaths/min)	3.9
VARC	Fraction arterial blood	0.025
VVC	Fraction venous blood	0.045
VALC	Fraction alveolar space	0.005
VLUC	Fraction lung tissue	0.006
VLIC	Fraction liver tissue	0.055
VFC	Fraction fat tissue	0.06
VKC	Fraction kidney tissue	0.017
VOC	Fraction "other" tissue	0.797
TCLU	Lung capillary volume (% of tissue volume)	11.0
TCLI	Liver capillary volume (% of tissue volume)	11.0
TCF	Fat capillary volume (% of tissue volume)	3.0
TCK	Kidney capillary volume (% of tissue volume)	10.2
TCO	"Other" capillary volume (% of tissue volume)	4.2
QLC	Fractional blood flow to liver (% of cardiac output)	0.162
QFC	Fractional blood flow to fat (% of cardiac output)	0.05
QKC	Fractional blood flow to kidney (% of cardiac output)	0.163
QOC	Fractional blood flow to "other" (% of cardiac output)	0.625
PF	Fat permeability	1.2
PO	"Other" permeability	2.7
Metabolic Parameters		
VMAXLI1	Capacity of saturable metabolism in liver (nmol/mL per minute)	229.6
KMLI1	Affinity of saturable metabolism in liver (nmol/mL)	40.2
VMAXLI2	Capacity of saturable metabolism in liver (nmol/mL per minute)	201.4
KMLI2	Affinity of saturable metabolism in liver (nmol/mL)	99.6
n	Hill constant	2
VMAXLU	Capacity of saturable metabolism in lung (nmol/mL per minute)	58.1
KMLU	Affinity of saturable metabolism in lung (nmol/mL)	40.2
Chemical Parameters		
PB	Blood/air partition coefficient	571
PLI	Liver/blood partition coefficient	7.0
PLU	Lung/blood partition coefficient	1.81
PF	Fat/blood partition coefficient	160.4
PK	Kidney/blood partition coefficient	4
PO	"Other"/blood partition coefficient	4
PERM	Capillary permeability constant	2.7
PERMF	Fat capillary permeability constant	1.2

TABLE 12.3 *(Continued)*
Parameters Used in Model Simulations

Parameter Symbol	Parameter Description	Parameter Value
Calculated Parameters		
$CO = HR*SV$	Cardiac output (L/min)	
$QLI = (QLC*1000)*CO$	Liver blood flow (ml/min)	
$QF = (QFC*1000)*CO;$	Fat blood flow (ml/min)	
$QK = (QKC*1000)*CO;$	Kidney blood flow (ml/min)	
$QO = (QOC*1000)*CO;$	"Other" blood flow (ml/min)	
$QTO = QLI+QF+QK+QO;$	Total blood flow (ml/min)	
$QP = Vt*f$	Alveolar ventilation (L/min)	
$QV = QP*1000$	Alveolar ventilation (ml/min)	
$VAR = (VARC*BW*1000)$	Arterial blood volume (ml)	
$VV = (VVC*BW*1000)$	Venous blood (ml)	
$VAL = (VALC*BW*1000)$	Alveolar space (ml)	
$VLU = (VLUC*BW*1000)$	Lung tissue (ml)	
$VLI = (VLIC*BW*1000)$	Liver tissue (ml)	
$VF = (VFC*BW*1000)$	Fat tissue (ml)	
$VK = (VKC*BW*1000)$	Kidney tissue (ml)	
$VO = (VOC*BW*1000)$	"Other" tissue (ml)	
$VCLU = ((TCLU/100)*VLU)$	Lung capillary volume (ml)	
$VCLI = ((TCLI/100)*VLI)$	Liver capillary volume (ml)	
$VCF = ((TCF/100)*VF)$	Fat capillary volume (ml)	
$VCK = ((TCK/100)*VK)$	Kidney capillary volume (ml)	
$VCO = ((TCO/100)*VO);$	"Other" tissue capillary volume (ml)	

Lung capillaries:

$$\frac{dAMT_{\text{lungcap}}}{dt} = \frac{AMT_{\text{ven}}}{V_{\text{ven}}} \cdot Q_{\text{total}} \cdot Perm + \frac{AMT_{\text{alv}}}{V_{\text{alv}}} Q_{\text{vent}} \cdot Perm + \tag{12.2}$$

$$\frac{AMT_{\text{lung}}}{V_{\text{lung}}} \cdot \frac{Q_{\text{total}}}{P_{\text{lung}}} \cdot Perm - \frac{AMT_{\text{lungcap}}}{V_{\text{lungcap}}} \cdot Q_{\text{total}} -$$

$$\frac{AMT_{\text{lungcap}}}{V_{\text{lungcap}}} \cdot Q_{\text{total}} \cdot Perm - \frac{AMT_{\text{lungcap}}}{V_{\text{lungcap}}} \cdot \frac{Q_{\text{vent}}}{P_{\text{air}}} \cdot Perm$$

Lung tissue:

$$\frac{dAMT_{\text{lung}}}{dt} = \frac{ANT_{\text{lungcap}}}{V_{\text{lungcap}}} \cdot Q_{\text{total}} \cdot Perm - \frac{AMT_{\text{lung}}}{V_{\text{lung}}} \cdot \frac{Q_{\text{total}}}{P_{\text{lung}}} \cdot Perm - \tag{12.3}$$

$$\frac{V_{\text{max lung}} \cdot V_{\text{lung}} \cdot AMT_{\text{lung}}}{K_{\text{mlung}} \cdot V_{\text{lung}} + AMT_{\text{lung}}}$$

Arterial blood:

$$\frac{dAMT_{\text{art}}}{dt} = \frac{AMT_{\text{lungcap}}}{V_{\text{lungcap}}} \cdot Q_{\text{total}} - \frac{AMT_{\text{art}}}{V_{\text{art}}} \cdot Q_{\text{total}} \tag{12.4}$$

Venous blood:

$$\frac{dAMT_{ven}}{dt} = \sum \frac{AMT_{tissuecap}}{V_{tissuecap}} \cdot Q_{tissue} - \frac{AMT_{ven}}{V_{ven}} \cdot Q_{total} \qquad (12.5)$$

The differential equations describing the liver and skin compartments include a term for metabolism of naphthalene. The equations for the liver and skin compartments include those for both capillaries and tissue.

Liver capillaries:

$$\frac{dAMT_{livercap}}{dt} = \frac{AMT_{art}}{V_{art}} \cdot Q_{liver} + \frac{AMT_{liver}}{V_{liver}} \cdot \frac{Q_{liver}}{P_{liver}} \cdot Perm - \\ \frac{AMT_{livercap}}{V_{livercap}} \cdot Q_{liver} - \frac{AMT_{livercap}}{V_{livercap}} \cdot Q_{liver} \cdot Perm \qquad (12.6)$$

Liver tissue:

$$\frac{dAMT_{liver}}{dt} = \frac{AMT_{livercap}}{V_{liver}} \cdot Q_{liver} \cdot Perm - \frac{AMT_{liver}}{V_{liver}} \cdot \frac{Q_{liver}}{P_{liver}} \cdot Perm - \\ \frac{V_{max\ liver1} \cdot V_{liver} \cdot AMT_{liver}}{K_{mliver1} \cdot V_{liver} + AMT_{liver}} - \frac{V_{max\ liver2} \cdot V_{liver} \cdot AMT_{liver}^{n}}{(K_{mliver2} \cdot V_{liver})^{n} + AMT_{liver}^{n}} \qquad (12.7)$$

Skin capillaries:

$$\frac{dAMT_{skincap}}{dt} = \frac{AMT_{art}}{V_{art}} \cdot Q_{skin} + \frac{AMT_{skin}}{V_{skin}} \cdot \frac{Q_{skin}}{P_{skin}} \cdot Perm - \ldots \\ \frac{AMT_{skincap}}{V_{skincap}} \cdot Q_{skin} - \frac{AMT_{skincap}}{V_{skincap}} \cdot Q_{skin} \cdot Perm \qquad (12.8)$$

Skin:

$$\frac{dAMT_{skin}}{dt} = \frac{AMT_{skincap}}{V_{skincap}} \cdot Q_{skin} \cdot Perm - \frac{AMT_{skin}}{V_{skin}} \cdot \frac{Q_{skin}}{P_{skin}} \cdot Perm - \ldots \\ \frac{V_{max\ skin} \cdot V_{skin} \cdot AMT_{skin}}{K_{mskin} \cdot V_{skin} + AMT_{skin}} \qquad (12.9)$$

Nonmetabolizing tissues: The differential equations for the other nonmetabolizing tissues and their capillaries in Figure 12.1, kidney, fat, and other, have the same structure.

$$\frac{dAMT_{tissuecap}}{dt} = \frac{AMT_{art}}{V_{art}} \cdot Q_{tissue} + \frac{AMT_{tissue}}{V_{tissue}} \cdot \frac{Q_{tissue}}{P_{tissue}} \cdot Perm - \\ \frac{AMT_{tissuecap}}{V_{tissuecap}} \cdot Q_{tissue} - \frac{AMT_{tissuecap}}{V_{tissuecap}} \cdot Q_{tissue} \cdot Perm \qquad (12.10)$$

$$\frac{dAMT_{tissue}}{dt} = \frac{AMT_{tissuecap}}{V_{tissuecap}} \cdot Q_{tissue} \cdot Perm - \frac{AMT_{tissue}}{V_{tissue}} \cdot \frac{Q_{tissue}}{P_{tissue}} \cdot Perm \qquad (12.11)$$

TABLE 12.4
Parameters for Simulating Porpoise Diving

| | Parameter | | |
Swim Mode	Heart Rate, *HR* Beats·min⁻¹	Breathing Frequency, *f* Breaths·min⁻¹	Sources
Predive	110	3.9	Irving et al. 1941
Dive	30	0	Meagher et al. 2002
			Williams et al. 1993
Postdive	100	10	Williams et al. 1999

Two equations are needed to model the effects of diving on ventilation (Q_{vent}) and cardiac output (*Co*) rates (L/min). The equation for ventilation rate is taken from Equation (5.7):

$$Q_{vent} = V_t \cdot f \tag{12.12}$$

where
V_t = tidal volume, L·breath⁻¹
f = respiration rate, breaths·min⁻¹

The equation for cardiac output rate is:

$$Q_{total} = Hr \cdot Sv \tag{12.13}$$

where
Hr = heart rate, beats·min⁻¹
Sv = stroke volume, L·beat⁻¹

12.4 DATA REQUIREMENTS

The parameter values used in the model are listed in Table 12.3. Most of these values are from the model of human exposure to naphthalene in JP-8 jet fuel (Dixon et al. 2001).

The pertinent bottlenose dolphin parameters, including those in Equations (12.12) and (12.13) were obtained from published literature. Bottlenose dolphin adult body weights cover a wide range with males significantly heavier than females. The mean body weight of four adult bottlenose dolphins reported by (Williams et al. 1999) was 197.5±17.8 kg (mean±S.D.). We used a value of 200 kg in the model. Williams et al. (1999) also reported predive heart rates of 101.8±0.7 beats·min⁻¹ for shallow dives and 111.3±2.2 beats·min⁻¹ for deeper dives. Meagher et al. (2002) reported heart rates that ranged from 72 to 101 beats·min⁻¹, whereas Sommer et al. (1968) found heart rates in five dolphins ranged from 84 to 140 beats·min⁻¹. We used a value of 110 beats·min⁻¹ for the predive value of parameter *Hr* (Table 12.4). During dives, bottlenose dolphins showed significant bradycardia, with heart rates as low as 37.0±1.8 beats·min⁻¹ for shallow dives and 30.0±2.2 beats·min⁻¹ for deeper dives (Williams et al. 1999). We used a value of 30 beats·min⁻¹ in the model. Williams et al. (1999) found no significant difference between predive and postdive heart rates, although the measured rates were 6–10% lower than predive levels. We used a value of 100 beats·min⁻¹ for the postdive rate.

Respiratory rates (breathing frequency) of bottlenose dolphins resting at the surface averaged 3.9±0.2 breaths·min⁻¹; respiratory rates following dives were 2.5 times that value (Williams et al. 1999). This higher rate is believed to be an adaptation to remove CO_2 stored during the dive (Boutilier et al. 2001). Respiration rates measured by Meagher et al. (2002) ranged between 2.3 and 3.5 breaths·min⁻¹ and those measured by Irving et al. (1941) ranged between 0.9 and 3.5 breaths·min⁻¹. Obviously, the respiratory rate during dives was zero.

12.5   MODEL VALIDATION

Because this model is a new model, there are little data available for testing. At this stage of development, we consider the model useful for conducting "what if" simulations of naphthalene uptake and distribution from inhalation using known parameter values for exposure scenarios related to the effects of diving behavior. Until all the parameter values have been measured or estimated, however, we can model the system and run simulations that can lead to hypotheses about how the system functions (see Section 1.2.4). We also conducted first-time model tests (see Section 10.1).

The random variable in the model is ambient naphthalene concentration. We tested for *internal validity* by varying both the mean and standard deviation of the concentrations. These tests included setting the standard deviation to zero for comparison with the stochastic output. *Face validity* was tested by having marine mammal experts review the model. We tested for *variable-parameter validity* by using parameter values obtained from studies on *T. truncatus* (Table 12.4). All the parameters used in the model have real-world counterparts (Table 12.3). Sensitivity tests also were conducted (see Section 12.7). The hypothesis that we were interested in formulating was that bottlenose dolphin diving behavior can influence the amounts of naphthalene in dolphin tissues by varying ventilation rates and cardiac output. A test of *hypothesis validity* shows a different response in naphthalene tissue amounts related to diving behavior. This hypothesis should be tested by measuring naphthalene amounts in *T. truncatus*. To test for *event validity*, such as the mortality in a population following exposure to naphthalene, requires a dose response to be determined experimentally.

12.6   DESIGN OF SIMULATION EXPERIMENTS

We conducted simulations for different exposure scenarios based on the sequence of ventilation and cardiac output rates. The first simulation was a constant exposure over 6 hours to see whether the naphthalene concentrations reached equilibrium. In this scenario, we assumed that the dolphins swam at the surface for the entire simulation. The second scenario was constant exposure for 30 minutes and no exposure for the next 2 hours to simulate the dolphins swimming out of the contaminated area to see if the dolphins were able to clear the naphthalene. The third scenario was constant exposure during swimming at the surface but at 10 minutes after initial exposure, the dolphins dived for 5 minutes (Figure 12.2). There also was a 5-minute recovery period, during which the breathing rate and heart rate were increased above normal surface swimming rates. This scenario examined the effects of not breathing and bradycardia during the dive. Of course, there was no exposure to naphthalene vapors during the dive. The model was designed to develop hypotheses about the effects of diving on naphthalene inhalation and distribution.

12.7   ANALYZE RESULTS OF SIMULATION EXPERIMENTS

12.7.1   SIMULATION OUTPUT

The results of the first scenario showed an initial rapid increase in the amounts of naphthalene in all tissues, for approximately one hour (Figure 12.3). In all but the fat and other compartments, this was followed by a more gradual increase over the next 5 hours. None of the compartments reached equilibrium.

In the second exposure scenario, there is a rapid increase in all compartments except fat and other, which showed more gradual increases during exposure (Figure 12.4). After the initial increase, naphthalene begins to decrease in all tissues except for fat and other tissues because of elimination. In fat and other tissues, the amounts of naphthalene continue to increase gradually because of depuration from the other tissues. The fat accumulation likely is a result of the high fat:blood partition coefficient. As in the other compartments, the predicted exhaled breath concentration increases rapidly after initial exposure and then drops rapidly after exposure.

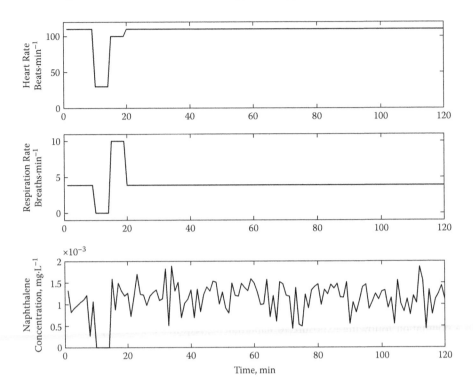

FIGURE 12.2 Exposure Scenario 3 of dolphins diving for 5 minutes after a 10-minute exposure to naphthalene vapors.

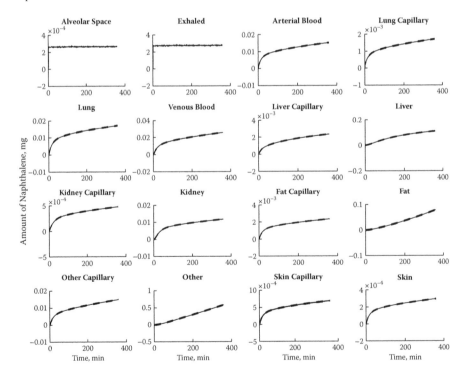

FIGURE 12.3 Predicted mean and 95% CI amounts of naphthalene (mg) in bottlenose dolphins for 6 hours of exposure.

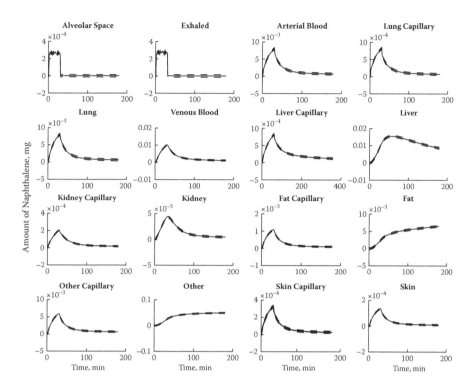

FIGURE 12.4   Predicted mean and 95% CI amounts of naphthalene (mg) in bottlenose dolphin tissues over 2½ hours following a 30-minute exposure while swimming at the surface.

In the third exposure scenario, predicted amounts of naphthalene in dolphin tissues start to increase rapidly during predive exposure (Figure 12.5). During the dives, naphthalene amounts drop in all tissues except liver, fat, and other tissues. The drop is most pronounced in capillaries, arterial blood, lungs, and skin. At the end of the dive, amounts again increase rapidly because of an increased respiration rate. After the recovery period, naphthalene amounts increase at a reduced rate, with patterns similar to those shown in Figure 12.3 for constant exposure.

The results in Figure 12.5 are for a simulation time span of 2 hours to show the effects of diving for 5 minutes. Time spans were extended to 6 hours so that these results can be compared with those from the constant exposures in Scenario 1 to assess the effects of diving behavior (Table 12.5).

The results of Scenario 2 show all compartments except fat and other approach zero after 2½ hours of elimination. We hypothesize that except for fat and other tissues, amounts of naphthalene would reach background levels. For fat and other tissues, the amounts remain high after 3 hours because of the relatively large percentage volumes of those tissues (0.21 and 0.63 for fat and other, respectively). The results from the comparison of Scenario 1, with constant exposure, and Scenario 3, with a single 5-minute dive, showed that diving behavior reduces the amount of naphthalene in tissues by an average of 9.57±3.00% except in fat, other, and skin compartments that increase by an average 6.27±3.02%.

12.7.2   SENSITIVITY ANALYSIS

We conducted sensitivity analyses, using Equation (9.8), of the noncapillary tissues to changes in several physiological parameters such as body weight (*BW*), heart rate (*HR*), and breathing frequency (*f*); metabolic parameters such as Affinity of Saturable Metabolism in the Lung (KMLU) and Capacity of Saturable Metabolism in the Lung (VMAXLU); and chemical parameters, such as Blood/Air Partition Coefficient (PB) and Capillary Permeability Constant (PERM). None of the

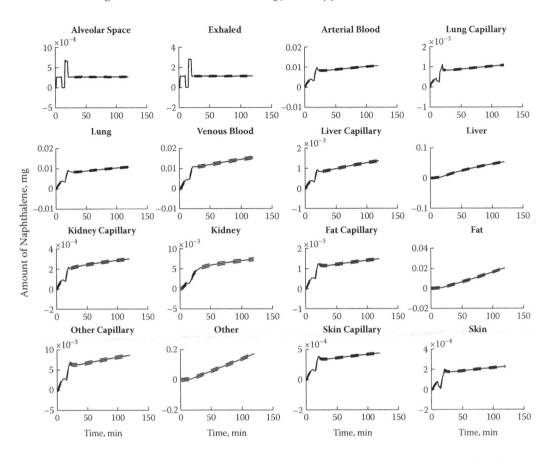

FIGURE 12.5 Predicted mean and 95% CI amounts of naphthalene (mg) in bottlenose dolphin tissues before and after a dive.

analyses identified "sensitive" parameters, that is, except for brief initial transients, the sensitivity, S, remained below | 1 |. We present the results only for VMAXLU in Figure 12.6, which are typical.

12.8 PRESENTATION AND IMPLEMENTATION OF RESULTS

We conducted simulations of bottlenose dolphins exposed to naphthalene vapors using a first-time model. The model can be used to form hypotheses relative to swimming behavior, particularly the effect of diving on naphthalene uptake. Naphthalene was chosen because it is a constituent of crude oil released in oil tanker spills or drilling platform blowouts and because we had some experience with a naphthalene PBTK model. The model could be used to simulate the uptake and distribution of other crude oil constituents, either singly or as mixtures, given the requisite parameter values. Before the model can be used to make predictions about the effects of exposure to crude oil, a dose–response function would have to be added to the model.

It is well known that exposure to naphthalene and other PAHs can cause neurological effects in human beings, mice, and rats. Because invasive research on marine mammals is restricted, it may be necessary to use parameter values from these studies and extrapolate to *T. truncatus* or other members of the Delphinidae. We also know that dolphin mortality has been associated with exposure to oil spill chemicals and some of these deaths occurred in nonoiled dolphins. It has been hypothesized that these deaths could be related to dolphins breathing toxic vapors. Whether the deaths were caused by a toxic response is not known. They may have been caused by an indirect effect of toxic chemicals interfering with the breathing mechanism.

TABLE 12.5
Results of simulations of three exposure scenarios. Values are amounts of naphthalene (mg) in bottlenose dolphin tissues.

Compartment	Scenario 1[a] 6 h Time Span	Scenario 2[b] 3 h Time Span	Scenario 3[c] 6 h Time Span	Percentage Change in Scenario 3 Compared with Scenario 1
Alveolar Space	2.69E–04	2.46E–07	2.64E–04	–2.02
Arterial Blood	1.52E–02	6.54E–04	1.50E–02	–1.32
Fat	7.84E–02	6.50E–03	8.46E–02	7.91
Fat Capillaries	2.30E–03	9.12E–05	2.10E–03	–8.70
Kidney	1.18E–02	4.79E–04	1.04E–02	–11.86
Kidney Capillaries	4.75E–04	1.87E–05	4.20E–04	–11.49
Liver	1.10E–01	8.60E–03	9.96E–02	–9.37
Liver Capillaries	2.30E–03	1.21E–04	2.10E–03	–8.70
Lung	1.71E–02	6.65E–04	1.52E–02	–11.11
Lung Capillaries	1.70E–03	6.57E–05	1.50E–03	–11.76
Other	5.82E–01	5.01E–02	5.98E–01	2.79
Other Capillaries	1.49E–02	6.89E–04	1.35E–02	–9.40
Skin	2.95E–04	1.01E–05	3.19E–04	8.13
Skin Capillaries	6.93E–04	6.93E–04	6.13E–04	–11.52
Venous Blood	2.57E–02	1.20E–03	2.31E–02	–10.12
Exhaled Breath	2.80E+00	2.60E–03	1.12E+00	–60.00

[a] Scenario 1 is constant exposure to naphthalene for 6 hours.
[b] Scenario 2 is constant exposure for 30 minutes, followed by zero exposure for 2½ hours.
[c] Scenario 3 is 10 minutes of constant exposure, 5 minutes of zero exposure, followed by 5¾ hours of exposure.

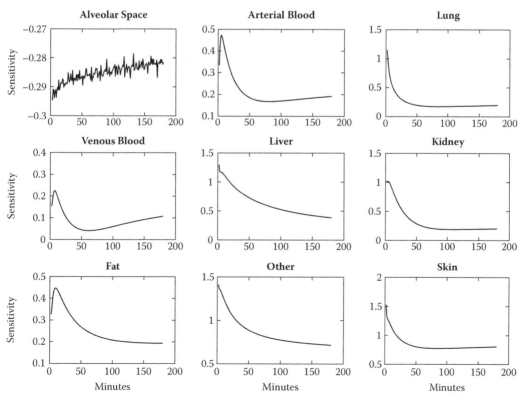

FIGURE 12.6 Sensitivity of naphthalene amounts in porpoise tissues to a 50% increase in the parameter.

The physiological mechanisms controlling the adaptation to asphyxia, necessary for diving, involve breathing, cardiac output, and tissue metabolism. Exposure to a toxic chemical could affect, directly or indirectly, any one or more of these mechanisms. It also may be possible that breathing toxic vapors simply prevents an adequate supply of oxygen, which also is needed for successful diving. These mechanisms should be added to the model to explore additional hypotheses. As we pointed out in Chapter 1, the combination of modeling and simulation together with field and laboratory studies should provide the best understanding of the effects of oils spills on marine mammals.

REFERENCES

Boutilier, R. G., J. Z. Reed, and M. A. Fedak. 2001. "Unsteady-State Gas Exchange and Storage in Diving Marine Mammals: The Harbor Porpoise and Gray Seal." *American Journal of Physiology Regulatory Integrative and Comparative Physiology* 281:R490–R494.

Dixon, K. R., E. P. Albers, and C. Chappell. 2001. "Risk Assessment: Modeling." International Conference on JP-8 Jet Fuel. San Antonio, Texas, August 7–10.

Irving, L., P. F. Scholander, and S. W. Grinnell. 1941. "The Respiration of the Porpoise, *Tursiops truncatus*. *Journal of Cellular and Comparative Physiology* 17:145–168.

ITOPF (The International Tanker Owners Pollution Federation Limited). 2010. *Oil Tanker Spill Statistics: 2009*. ITOPF, London. http://www.itopf.com/information-services/data-and-statistics/statistics/documents/Statspack2009-FINAL.pdf (accessed 5/22/11).

Meagher, E. M., W. A. McLellan, A. J. Westgate, R. S. Wells, D. Frierson Jr., and D. A. Pabst. 2002. "The Relationship between Heat Flow and Vasculature in the Dorsal Fin of Wild Bottlenose Dolphins *Tursiops truncatus*." *Journal of Experimental Biology* 205:3475–3486.

Quick, D. J., and M. L. Shuler. 1999. "Use of *In Vitro* Data for Construction of a Physiologically Based Pharmacokinetic Model for Naphthalene in Rats and Mice to Probe Species Differences." *Biotechnology Progress* 15:540–555.

Ramsey, J. C., and M. E. Andersen. 1984. "A Physiologically Based Description of the Inhalation Pharmacokinetics of Styrene in Rats and Humans." *Toxicology and Applied Pharmacology* 73:159–175.

Robinson, P. J. 2000. "Pharmacokinetic Modeling of JP-8 Jet Fuel Components. I. Nonane and C9-C12 Aliphatic Components." ManTEch–GEO-CENTERS Joint Venture, Dayton, OH.

Sommer, L. S., W. L. McFarland, R. E. Galliano, E. L. Nagel, and P. J. Morgane. 1968. "Hemodynamic and Coronary Angiographic Studies in the Bottlenose Dolphin (*Tursiops truncatus*)." *American Journal of Physiology* 215:1498–1505.

Sweeney, L. M., M. L. Shuler, D. J. Quick, and J. G. Babish. 1996. "A Preliminary Physiologically Based Pharmacokinetic Model for Naphthalene and Naphthalene Oxide in Mice and Rats." *Annals of Biomedical Engineering* 24:305–320.

U.S. Department of Health and Human Services. 2000. *NTP Technical Report on the Toxicology and Carcinogenesis Studies of Naphthalene (CAS No. 91-20-3) in F344/N Rats (Inhalation Studies)*. National Toxicology Program, NTP TR 500, NIH Publication No. 01-4434. Research Triangle Park, NC: National Toxicology Program.

Williams, T. M., W. A. Friedl, and J. E. Haun. 1993. "The Physiology of Bottlenose Dolphins (*Tursiops truncatus*): Heart Rate, Metabolic Rate and Plasma Lactate Concentration during Exercise." *Journal of Experimental Biology* 179:31–46.

Williams, T. M., J. E. Haun, and W. A. Friedl. 1999. "The Diving Physiology of Bottlenose Dolphins (*Tursiops truncatus*) I. Balancing the Demands of Exercise for Energy Conservation at Depth." *Journal of Experimental Biology* 202:2739–2748.

13 Case Study
Simulating the Effects of Temperature Plumes on the Uptake of Mercury in Daphnia

13.1 PROBLEM DEFINITION

This problem is to explore the effects of simulated thermal plumes on mercury dynamics in the zooplankter *Daphnia pulex* based on data from Huckabee et al. (1977) and Dixon (1977). A model was developed to simulate the uptake and elimination of mercury in *Daphnia* as a function of temperature. The model was used to simulate mercury dynamics in *Daphnia* in thermal plumes resulting from power plant cooling water discharge (Figure 13.1). Water temperature in the discharge stream will be determined largely by the discharge velocity. Because *Daphnia* move passively in a plume, the temperature profile of the plume should define the temperatures to which the *Daphnia* are exposed. Where other water temperature profiles can be defined, the model should be able to predict mercury dynamics under those conditions as well.

13.2 MODEL DEVELOPMENT

A conceptual model is relatively simple with mercury concentrations in aggregated populations of *Daphnia* being the only state variable, and ambient mercury concentrations and temperature as control variables. The modeling approach uses ordinary differential equations to describe the mercury dynamics. The model parameters defining the uptake and elimination rates are hypothesized to change with temperature.

A logical flow diagram shows the flow of mercury between *Daphnia* and water (Figure 13.2).

13.3 MODEL IMPLEMENTATION

Experimental data (Figure 13.3 and Figure 13.4) suggest that elimination and uptake can be modeled using Equations 3 and 4, respectively, in Table 2.1. A slight modification to the elimination equation includes a parameter for a minimum level of mercury concentration as well as a maximum level. The expression for elimination, Q_e, is (see Example 2.2):

$$Q_e = (p_5 - p_4) e^{-p_2 t} + p_4 \qquad (13.1)$$

where
  p_5 = the maximum (initial) mercury concentration in elimination experiments
  p_4 = the minimum (final) mercury concentration in elimination experiments
  p_2 = elimination rate constant

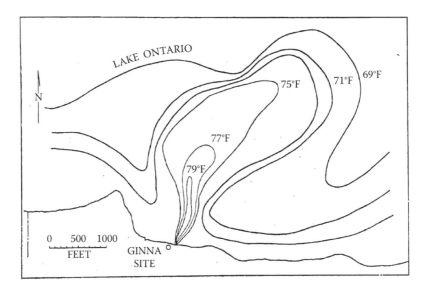

FIGURE 13.1   A thermal plume from a power plant located on Lake Ontario.

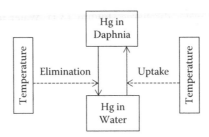

FIGURE 13.2   Flow diagram for mercury dynamics in *Daphnia pulex*.

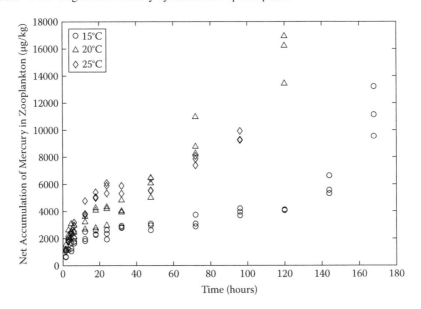

FIGURE 13.3   Mercury concentrations (µg/kg) in *Daphnia pulex* during uptake experiments.

The minimum concentration during uptake was assumed to be zero at time zero, so the expression for uptake, Q_u, is:

$$Q_u = (p_1)(1 - e^{-p_3 t}) \qquad (13.2)$$

where

p_1 = the maximum (final) mercury concentration in uptake experiments
p_3 = uptake rate constant

13.4   DATA REQUIREMENTS

Because the purpose of this model is to explore alternatives of temperature regimes in thermal plumes and their relative impact on mercury dynamics in *Daphnia pulex*, data are required to parameterize the uptake and elimination terms in the model. Initial values of mercury body burden should be at background levels. Independent data from field studies are not required because accurate predictions are not a goal, only relative mercury concentrations in *Daphnia* exposed to different thermal plumes. Because water temperature in the discharge stream will be determined largely by the discharge velocity, and to simplify the comparison of alternatives, only discharge velocity is considered as a perturbation to the system.

The data come from experiments conducted by Huckabee et al. (1977). The model in this case study is somewhat different from that in Dixon (1977). The data were analyzed in the following sequence to estimate the model parameters.

1. The data were plotted.
2. Outliers were removed and the edited data were plotted.
3. Parameters were estimated by fitting Equations (13.1) and (13.2) to the censored data using nonlinear regression.
4. A model of gross uptake rate was formed by adding the elimination rate to the net uptake rate.
5. Parameters for gross uptake were estimated using nonlinear regression.
6. Differential equations were developed for mercury dynamics from the gross uptake and elimination rates.
7. Model parameters were written as functions of temperature.
8. The temperatures in thermal plumes were estimated.

13.4.1   PLOT DATA

Both the uptake and elimination experiments included three replicate measures of mercury concentration at each time (Figures 13.3 and 13.4).

These data illustrate the importance of understanding the mechanisms involved in measuring concentrations in uptake and elimination experiments. After approaching an asymptote at about 72 hours, the mercury concentration begins to increase exponentially in the uptake experiments. This result was caused by reduced weight of the *Daphnia*, as they were not fed during the experiments. In the case of elimination, experiments at 15°C and 20°C also showed an increase in concentration. This phenomenon has been observed in elimination experiments where different animals are sampled over time. We know, however, that in the absence of additional inputs of mercury, there has to be monotonically decreasing concentrations over time. In other words, each successive concentration has to be less than the preceding concentration.

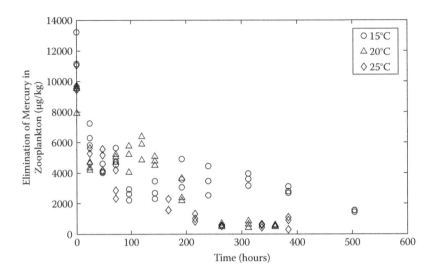

FIGURE 13.4   Mercury concentrations (μg/kg) in *Daphnia* during elimination experiments.

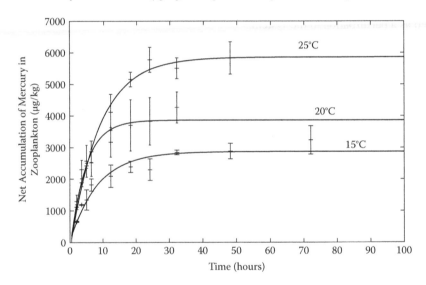

FIGURE 13.5   Net accumulation of mercury in *Daphnia pulex* using edited data. (Adapted from K. R. Dixon. 1977. "Thermal Plumes and Mercury Dynamics in Zooplankton." In *International Conference on Heavy Metals in the Environment, Vol. 2*, 875–886, Toronto, Ontario, Canada.)

13.4.2   PLOT EDITED DATA

We edited the data to account for these factors and plotted the resulting means and standard deviations for uptake and elimination in Figures 13.5 and 13.6, respectively. The elimination data were also expressed as the percentage mercury remaining to facilitate comparison among temperatures.

13.4.3   ESTIMATE MODEL PARAMETERS

We used the MATLAB function **nlinfit** to fit the models to the data (see Section 7.2.1). The parameter values are in a vector **betahat**. Other statistics generated in the output in the Command Window include the residuals, **r**, covariance matrix, **sigma**, confidence intervals for

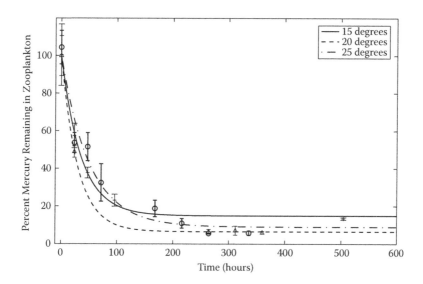

FIGURE 13.6 Elimination of mercury in *Daphnia* using edited data. (Adapted from K. R. Dixon. 1977. "Thermal Plumes and Mercury Dynamics in Zooplankton." In *International Conference on Heavy Metals in the Environment, Vol. 2*, 875–886, Toronto, Ontario, Canada.)

TABLE 13.1
Uptake and Elimination Model Parameters
Estimated Using Nonlinear Regression

Temperature °C	Parameter Value				
	P1	P2	P3	P4	P5
15	2809.0	0.04167	0.1437	1676.8	11248.4
20	3828.3	0.03292	0.2335	586.7	9029.0
25	5828.9	0.01950	0.1171	864.3	9638.4

the parameters, **betaci**, the predicted *y* values, **yhat**, and the half-confidence intervals on the predicted values, **delta**. The following statements generate the output. Complete MATLAB programs are included on the enclosed CD: **daphnia_mercury15**, **daphnia_mercury20**, and **daphnia_mercury25**.

```
[betahat,r,sigma] = nlinfit(allday,allconc,@daphnia_mercury,beta)
betaci = nlparci(betahat,r,'covar',sigma)
[yhat, delta] = nlpredci(@daphnia_mercury,allday,betahat,r,sigma)
```

All the parameters in the uptake and elimination equations were estimated using the same nonlinear regression procedure. The resulting parameter estimates are included in Table 13.1.

13.4.4 GROSS UPTAKE MODEL

Because *Daphnia* excrete mercury at the same time that they are accumulating it, the uptake experiments measure net uptake. To estimate the true gross uptake, the elimination rate can be added to the uptake rate. This is done with the derivative forms of Equations (13.1) and (13.2). The model is written by summing the two rates in a differential equation:

$$\frac{dQ_{u*}}{dt} = p_3(p_1 - Q) + p_2 Q$$

$$= p_3 p_1 - (p_3 - p_2)Q$$

(13.3)

where Q_{u*} is the gross uptake rate and the other parameters were defined previously. The following m-files yield the solution to Equation (13.3) at ten time steps in the vector **GEN**.

```
%grossuptake3
%program to solve gross uptake equation in function grossuptake1
y0=0;
tspan=[0.0 100.0];
GEN = [];
y1 = [];

sol = ode45(@grossuptake1,tspan,y0);
   x = linspace(0, 100,10);
   y1 = deval(sol,x,1);
   GEN = [GEN;y1]

plot(x,GEN)
xlabel ('Time');
ylabel('Gross Uptake');
grid;

%grossuptake1
%program to model "gross uptake", i.e. net uptake plus elimination
function ydot=grossuptake1(~,y)
%p1 = 5828.9;          %Maximum concentration 25 degrees
%p1 = 3828.3;                            %20 degrees
p1 = 2809.0;                            %15 degrees
%p2 = 0.01950;         %Elimination rate constant 25 degrees
%p2 = 0.03292                           %20 degrees
p2 = 0.04167;                           %15 degrees
%p3 = 0.1171;          %Uptake rate constant 25 degrees
%p3 = 0.2335;                           %20 degrees
p3 = 0.1437;                           %15 degrees

ydot=p3*p1-y*(p3-p2);
```

The model is solved for the three temperatures. Parameter estimates for the new uptake model were obtained by fitting Equation (13.2) to the ten values of gross uptake.

13.4.5 ESTIMATE PARAMETERS FOR GROSS UPTAKE MODEL

The following m-files **fitted_mercury_uptake15** and **function daphnia_mercury**, were used to estimate the new model parameters at 15°C using the MATLAB nonlinear regression function, **nlinfit**. This program and the m-files for 20°C and 25°C are included on the enclosed CD.

```
%program fitted_mercury_uptake15
format long
```

```
conc = [0
    2.683026948905384
    3.546292251476606
    3.824223308695165
    3.913704018171389
    3.942523082531268
    3.951806218840622
    3.954797690699898
    3.955762244853819
    3.956073479272539];
conc = conc*1.0e+003;

day = [0
    0.111111111111111
    0.222222222222222
    0.333333333333333
    0.444444444444444
    0.555555555555556
    0.666666666666667
    0.777777777777778
    0.888888888888889
    1.000000000000000];

day = day*1.0e+002;

plot(day, conc,'ko')
xlabel('Day')
ylabel('Concentration, ppb')

beta = [5000; .1144];
yhat = [];
newx = 0:1:250;

[betahat,r,sigma] = nlinfit(day,conc,@daphnia_mercury,beta)
betaci = nlparci(betahat,r,'covar',sigma)
[yhat, delta] = nlpredci(@daphnia_mercury,day,betahat,r,sigma)
ucl = yhat + delta;
lcl = yhat - delta;

figure
plot(day, yhat,'k-')
hold on
plot(day, ucl, 'k--')
plot(day, lcl, 'k--')
xlabel('Day')
ylabel('Hg Concentration, \mug/kg')
title('15 Degrees')

%function to model uptake and elimination
function yhat = daphnia_mercury(beta,day)
```

TABLE 13.2
Gross Uptake and Elimination Model Parameters
Estimated Using Nonlinear Regression

Temperature	Parameter Value				
°C	P1*	P2	P3*	P4	P5
15	3956.2	0.04167	0.1020	1676.8	11248.4
20	4456.5	0.03292	0.2006	586.7	9029.0
25	6993.5	0.01950	0.0976	864.3	9638.4

Source: Adapted from the U.S. Department of Health and Human Services. 2000.

```
p1 = beta(1);
p3 = beta(2);
%p4 = beta(3);

yhat = p1*(1-exp(-p3*day));          %uptake
%yhat = (p5-p4)*(exp(-p2*day))+p4;    %elimination
```

The resulting set of parameters is shown in Table 13.2.

13.4.6   DIFFERENTIAL EQUATION FOR MERCURY DYNAMICS

We now can write the differential equation for the model of mercury dynamics that includes terms for gross uptake and elimination:

$$\frac{dQ}{dt} = p_3^* \ (p_1^* - Q) - p_2 Q \tag{13.4}$$

where p_1^* and p_3^* are the gross uptake parameters. Before we can run simulations, we have to write the parameter values as functions of temperature.

13.4.7   PARAMETERS AS FUNCTIONS OF TEMPERATURE

As we have described the procedures for model development, the first step is to plot the data, in this case parameter values, against temperature (Figure 13.7). The relation between p_1 and temperature we assumed to be piecewise linear. The relation between p_2 and temperature closely approximates a straight line, so we used simple linear regression to fit a line to the data. The parameter p_3 appears to show a triangular distribution. We calculated a piecewise linear relationship with this parameter also.

The function for p_1 is:

$$p_1 = \begin{cases} 2454.8 + 100.1 * temp & temp \leq 20 \\ -5691.4 + 507.4 * temp & temp > 20 \end{cases} \tag{13.5}$$

The function for p_2 is:

$$p_2 = 0.07570 - 0.002217 * temp \tag{13.6}$$

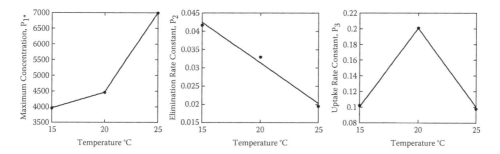

FIGURE 13.7   Parameters p_1, p_2, and p_3 as functions of temperature.

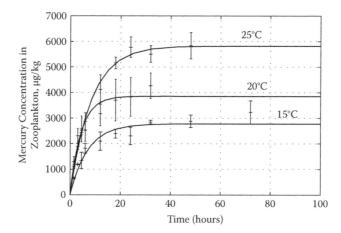

FIGURE 13.8   Net accumulation of mercury in *Daphnia pulex* with parameters written as functions of temperature.

The function for p_3 is:

$$p_3 = \begin{cases} -0.20 + 0.02 * \text{temp} & \text{temp} \le 20 \\ 0.60 - 0.02 * \text{temp} & \text{temp} > 20 \end{cases} \tag{13.7}$$

At this point, we tested the temperature parameter functions by including them in the model of mercury dynamics. The resulting plots of net uptake are shown in Figure 13.8. This plot is similar to the plots of net uptake at each of the individual temperatures (Figure 13.5), an indication of internal validity of the parameter functions.

13.4.8   ESTIMATE THERMAL PLUME TEMPERATURES

Before simulating mercury dynamics in *Daphnia*, we need to generate temperature profiles in thermal plumes. We used temperature data from a thermal plume from a real-world nuclear power plant, and temperatures from predicted isotherms (Rochester Gas & Electric [RG&E] 1974), to plot the increase in temperature above ambient water temperature (ΔT °C) at the center of the plume. Both the real-world plume and the simulated plumes showed a monotonically decreasing ΔT with distance from the point of discharge. Figure 13.9 shows the simulated plume ΔT for both average and low surface cooling obtained from the m-file **plotDeltaTvsTime.m**.

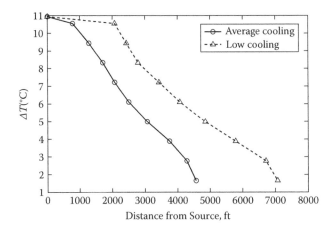

FIGURE 13.9   Increased temperatures above ambient water temperature (ΔT °C) at increasing distance from the point of discharge. Dashed line is for low surface cooling conditions and the solid line for average surface cooling conditions. (Data from Rochester Gas and Electric Co., "Environmental Report, Construction Permit Stage, Sterling Power Project Nuclear Unit No. 1." Docket No. STN-50-485 (1974).)

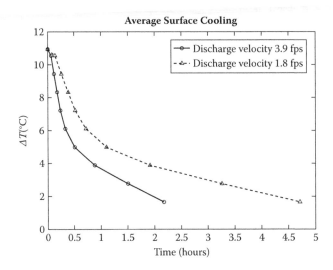

FIGURE 13.10   Increased temperatures above ambient water temperature (ΔT °C) at increasing time from the time of discharge for average surface cooling conditions. Dashed line is for a discharge velocity of 1.8 fps and the solid line for a discharge velocity of 3.9 fps. (Adapted from K. R. Dixon. 1977. "Thermal Plumes and Mercury Dynamics in Zooplankton." In *International Conference on Heavy Metals in the Environment, Vol. 2*, 875–886, Toronto, Ontario, Canada.)

We could simulate ΔT spatially using partial derivatives (Section 2.2.1). To simplify the model, we converted the ΔT data to a function of time by dividing the distance by the rate of flow in the plume. The rate of flow at the point of discharge is the discharge rate. We assumed a linear decrease in flow rate until the ΔT was zero at the ambient current of 7% of the discharge velocity (RG&E 1974). The resulting temperature profiles for average surface cooling conditions and low surface cooling conditions at two dicharge flow rates, 3.9 and 1.8 fps, are shown in Figures 13.10 and 13.11, respectively. See m-file **isotherms.m** for the code to generate this figure.

The second step in generating the thermal plume temperature model was to fit a model to the data in Figures 13.10 and 13.11. We fitted the half-Gaussian model (Equation [13.8]) that was used in

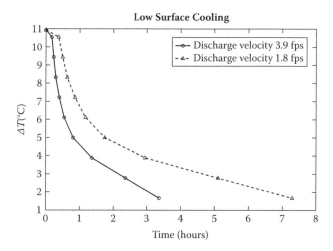

FIGURE 13.11 Increased temperatures above ambient water temperature (ΔT °C) at increasing time from the time of discharge for average surface cooling conditions. Dashed line is for a discharge velocity of 1.8 fps and the solid line for a discharge velocity of 3.9 fps. (Data from Rochester Gas and Electric Co., "Environmental Report, Construction Permit Stage, Sterling Power Project Nuclear Unit No. 1." Docket No. STN-50-485 (1974).)

Section 5.1.2, using the MATLAB nonlinear regression function **nlinfit** in the function **nlinregr** and m-file **nlinregr2**:

$$\Delta T = b_0 e^{(-b_1 t^{b_2})} \tag{13.8}$$

The parameter values from the regression were $b_0 = 11.44$, $b_1 = 1.19$, and $b_2 = 0.735$ for the 3.9 fps discharge velocity and $b_0 = 11.44$, $b_1 = 0.673$, and $b_2 = 0.734$ for the 1.8 fps discharge velocity. We now can generate temperature profiles with the m-file **plotDeltaTvsTime.m** (Figures 13.12 and 13.13).

13.5 MODEL VALIDATION

The purpose of the model also determines the level of validation. In this case study, we are not using the model to make predictions about the mercury dynamics in *Daphnia pulex* exposed to thermal plumes from a specific power plant. We are interested primarily in comparing the relative mercury dynamics caused by simulated thermal plumes with different discharge velocities, ambient water temperatures, and different surface cooling conditions. The validation methods we used then are those considered for a first-time model (Section 10.2). The internal validity of the model was demonstrated by the ability of the complete model, that is, where parameters are written as a function of temperature, to mimic the experimental data.

13.6 DESIGN OF SIMULATION EXPERIMENTS

The design of simulation experiments in this case study was intended to explore the effects of several variables on the mercury dynamics in *Daphnia pulex*. The controlling variables are the ambient water temperatures, the surface cooling conditions, and the velocity of the thermal discharge from the power plant cooling system. The objective in this design of a simulation experiment is to gain knowledge about the relationship between the thermal profiles of a thermal plume and the

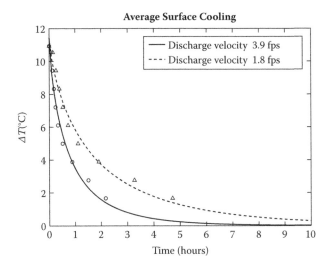

FIGURE 13.12   Observed and predicted thermal profile in a thermal plume with average surface cooling conditions. Dashed line is for a discharge velocity of 1.8 fps and the solid line for a discharge velocity of 3.9 fps. (Data from Rochester Gas and Electric Co., "Environmental Report, Construction Permit Stage, Sterling Power Project Nuclear Unit No. 1." Docket No. STN-50-485 (1974).)

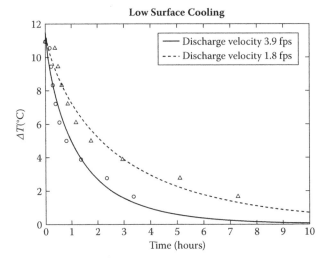

FIGURE 13.13   Observed and predicted thermal profile in a thermal plume with low surface cooling conditions. Dashed line is for a discharge velocity of 1.8 fps and the solid line for a discharge velocity of 3.9 fps. (Data from Rochester Gas and Electric Co., "Environmental Report, Construction Permit Stage, Sterling Power Project Nuclear Unit No. 1." Docket No. STN-50-485 (1974).)

mercury dynamics in *Daphnia pulex*. We assume a background level of mercury concentration for given ambient water temperatures. The input disturbance is the thermal plume temperature profile, and the output is the simulated state variable, mercury concentration in *Daphnia pulex*. The experimental design includes two levels of each factor, ambient water temperatures, surface cooling conditions, and discharge velocity, for a 2^3 factorial design. Ambient water temperatures are 15 and 20°C, surface cooling conditions are classified as average or low, and discharge velocities are 1.8 and 3.9 feet per second (fps). Initial values of mercury concentration in the water under ambient conditions were determined by running simulations at constant temperatures of 15 and 20°C.

13.7 ANALYZE RESULTS OF SIMULATION EXPERIMENTS

The analysis of these simulation experiments can explain the relative importance of the different factors affecting mercury dynamics in *Daphnia* exposed to thermal plumes. We used several measures of system stability (Section 9.2.1) to compare the effects of different factors, including (1) the maximum mercury concentration in *Daphnia* in the plumes, (2) the time to reach the maximum, (3) the overshoot, and (4) the settling time. We defined the overshoot as the difference between the equilibrium mercury concentration and the maximum value during the simulated thermal plume exposure. The settling time was defined as the time it took for the mercury concentration in the *Daphnia* to reach 5% of the equilibrium concentration.

13.8 PRESENTATION AND IMPLEMENTATION OF RESULTS

Results of the simulations are shown in Figures 13.14 through 13.17 and Table 13.3. In these results, mercury concentration increases rapidly as the zooplankton enter the plume. The concentration reaches a maximum within a few hours and then decreases exponentially. In general, concentrations at ambient water temperatures of 20°C are higher than at 15°C. The maximum concentration reached in any simulation was 4531.1 μg/kg at 20°C with low surface cooling conditions and a discharge velocity of 1.8 fps. Low surface cooling tended to result in higher concentrations more than did average conditions. Lower discharge velocities mean that zooplankton remain in the plume longer, and therefore are exposed to elevated temperatures over a longer period of time, compared with higher discharge velocities.

Other factors showed a slightly different pattern from maximum concentration. The highest overshoot was 678.4 μg/kg at 20°C with low surface cooling conditions and a discharge velocity of 1.8 fps. The second highest (672.3 μg/kg) was at 15°C, with low cooling, and 1.8 fps. The time

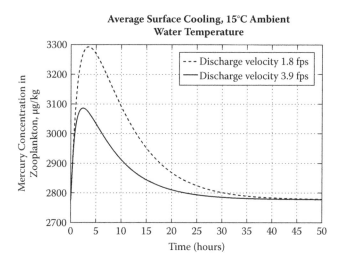

FIGURE 13.14 Simulated net mercury uptake in a thermal plume with average surface cooling conditions at 15°C ambient water temperature. (Adapted from K. R. Dixon. 1977. "Thermal Plumes and Mercury Dynamics in Zooplankton." In *International Conference on Heavy Metals in the Environment, Vol. 2*, 875–886, Toronto, Ontario, Canada.)

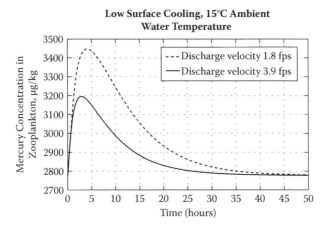

FIGURE 13.15 Simulated net mercury uptake in a thermal plume with low surface cooling conditions at 15°C ambient water temperature.

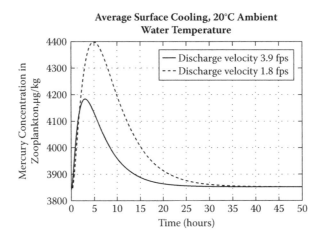

FIGURE 13.16 Simulated net mercury uptake in a thermal plume with average surface cooling conditions at 20°C ambient water temperature. (Adapted from K. R. Dixon. 1977. "Thermal Plumes and Mercury Dynamics in Zooplankton." In *International Conference on Heavy Metals in the Environment, Vol. 2*, 875–886, Toronto, Ontario, Canada.)

to maximum concentration was consistently higher at 20°C than at 15°C. Again, the conditions that produced the highest overshoot were 20°C, low cooling, and 1.8 fps. The same conditions also yielded the longest time to reach maximum concentration, at 6.7 hours. The conditions that resulted in the shortest time (2.5 hours) were the same as those that produced the lowest maximum concentration and overshoot: 15°C, average cooling, and 3.9 fps. The longest settling times were at 15°C compared to 20°C. The longest settling time was 32.6 hours at 15°C, with low cooling, and 1.8 fps. The shortest settling time was 18.4 hours at 20°C, with average cooling, and 3.9 fps, almost half of the longest settling time.

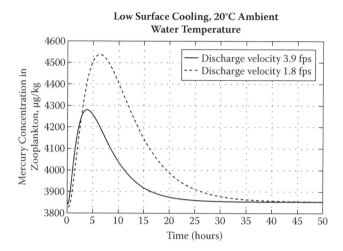

FIGURE 13.17   Simulated net mercury uptake in a thermal plume with low surface cooling conditions at 20°C ambient water temperature.

TABLE 13.3
Simulated Mercury Dynamics in Zooplankton Output Analysis

Ambient Water Temperature		Average Surface Cooling Conditions		Low Surface Cooling Conditions	
		1.8 fps	3.9 fps	1.8 fps	3.9 fps
15°C	Equilibrium concentration (µg/kg)	2775.5	2775.5	2775.5	2775.5
	Maximum concentration (µg/kg)	3286.4	3076.6	3447.8	3195.0
	Overshoot (µg/kg)	510.9	301.1	672.3	419.5
	Time to maximum (h)	3.9	2.5	4.0	3.2
	Settling time (h)	30.0	26.3	32.6	27.2
20°C	Equilibrium concentration (µg/kg)	3852.7	3852.7	3852.7	3852.7
	Maximum concentration (µg/kg)	4389.7	4166.7	4531.1	4271.7
	Overshoot (µg/kg)	537.0	314.0	678.4	419.0
	Time to maximum (h)	4.8	3.1	6.7	3.9
	Settling time (h)	24.3	18.4	28.2	20.2

REFERENCES

Dixon, K. R. 1977. "Thermal Plumes and Mercury Dynamics in Zooplankton." In *International Conference on Heavy Metals in the Environment, Vol. 2*, 875–886. Toronto, Ontario, Canada.

Huckabee, J. W., R. A. Goldstein, S. A. Janzen, and S. E. Woock. 1977. "Methylmercury in a Freshwater Foodchain." In *International Conference on Heavy Metals in the Environment, Vol. 2*, 199–216. Toronto, Ontario, Canada.

Rochester Gas and Electric Co. (RG&E). 1974. "Environmental Report. Construction Permit Stage, Sterling Power Project Nuclear Unit No. 1." Docket No. STN-50-485.

U.S. Department of Health and Human Services. 2000. *NTP Technical Report on the Toxicology and Carcinogenesis Studies of Naphthalene (CAS No. 91-20-3) in F344/N Rats (Inhalation Studies)*. National Toxicology Program, NTP TR 500, NIH Publication No. 01-4434. Research Triangle Park, NC: National Toxicology Program.

Index

Printed and bound by CPI Group (UK) Ltd, Croydon, CR0 4YY

21/10/2024

01777095-0014